INTRODUCTION

Now for the first time, we have compiled an accurate record of the features of each **Pz.Kpfw.38(t) Ausfuehrung** from **A** to **G** plus **S**. Previous attempts to fit this puzzle together correctly failed due to either missing or erroneous pieces of the puzzle, including:
o misidentifications due to the lack of **Pz.Kpfw.38(t)** identified by **Fgst.Nr.** (chassis numbers) resulting in the **Ausf.D** previously being confused with the **Ausf.S**,
o the hull height drawn over 30 mm too high on an original 1:5 scale overview drawing of the **Pz.Kpfw.38(t)** chassis (including copies in manuals, see page 18-3),
o an insufficient number of wartime photos to spot the numerous minor differences between **Ausfuehrung**, and
o not taking the research trips necessary to thoroughly examine and measure survivors in detail.

All of these deficiencies have been eliminated by a lot of hard work, research trips to measure and study the survivors, close attention to detail, and help from friends sharing their photo collections as well as photographing and measuring various models of turrets in Norway.

Over one thousand hours were spent carefully measuring and drawing all the details of surviving **Pz.Kpfw.38(t)** to create accurate as-built drawings. In addition, over two hundred "clean" photos taken during the war were carefully examined to determine precise details of bits now missing from the "survivors" and to determine the exact order in which modifications were introduced into the production series. Extra efforts were also expended to discover and accurately measure details on the chassis and belly which normally remain hidden from view even on the best quality wartime photos. It is this careful and thorough study of details, combined with over 38 years spent searching for surviving documents, drawings, and photos, that makes Panzer Tracts the ultimate in historically accurate documentation.

Measured and drawn at 1:1, the drawings are printed at the popular 1:35 scale. Even so, there is no loss in detail with this reduction in scale, because the software now being used to create the printing plates draws the lines, circles, arcs, hexes, and ovals one at a time - just like we did in creating these accurate as-built drawings.

As is our high standard, Panzer Tracts are based solely on surviving specimens, wartime photographs, and the content of primary source documents written by those who participated in the design, production, and employment of the Panzers. The real value of the **Pz.Kpfw.38(t)** can be learned by reading the 18 pages of operational history, including the translated experience reports written by unit commanders close to the time when the actions occurred.

Excerpts from the original operating manuals have been translated to provide a functional description of the **Pz.Kpfw.38(t)**, including lesser known features like the **Signalkasten** (red, blue, and green lights for communication between the commander and driver), which was a much more reliable device than an intercom system. Photos and drawings from these operating manuals have also been included to illustrate the interior of the **Pz.Kpfw.38(t)**. We were surprised to discover that the interior photos in manual D652/38 dated 1Sep43 were taken inside a **Slovak LT-38** (closely patterned after an **Ausf.D** but without the radio sets and thicker front armor) instead of a **Pz.Kpfw.38(t)**!

Long before Germany occupied Czechoslovakia in March 1939, the Czech company CKD had designed a unique series of inexpensive tanks for export, including the AH-IV and TNH for Iran, AH-IV S for Sweden, LTL-P for Peru, and LTL-H for Switzerland. Unlike the Skoda-designed S II a (LT.Sk.35), which was merely a knock-off copy of a 1932 vintage Vickers 6 ton export tank with complicated pneumatic controls added, CKD had been innovative in improving the reliability, drive ability, and maneuverability of their series of light tanks.

CKD was busy assembling the first series of 150 TNH-P light tanks that had been ordered for the Czech army when Germany occupied Bohemia and Moravia on 15 March 1939. Needing additional gun-armed light tanks, the Heeres Waffenamt awarded a contract to B.M.M. (Boehmisch-Maehrische Maschinenfabrik - the German wartime designation for CKD) to complete production of the first series of 150 **Panzerkampfwagen L.T.M 38**. Contract extensions then resulted in the **Pz.Kpfw.38(t) Ausf.B** to **G**, and a confiscated foreign order resulted in the **Ausf.S**.

As with all German weapons developed over a period of several years, the names evolved with time. To avoid confusion created by using inexact postwar popular names, this history maintains the correct names for each period as they were used in the original documents.

When first accepted by the Waffenamt, their production statistics were reported by Wa J Rue 6 from May through August 1939 as being **tschechische Pz.Kpfw.III**.

On 12 August 1939, A.O.K.8 reported that the repair of the **Panzerkampfwagen (3.7 cm) L.T.M 38** would be the responsibility of the firm Ceskomoravska Kolben-Danek A.G., Werk Lieben near Prague. The abbreviation **L.T.M 38** stood for **Leichte Tank Modell 38**.

From 13 October 1939 to 16 January 1940, Wa J Rue 6 referred to these **Panzerkampfwagen** as LTM 38 **Protektorat**.

On 16 January 1940 AHA/AgK/In6 announced that the names of the **tschechischen Panzerkampfwagen** taken over by the Panzertruppen previously known by the Czech name **Panzerkampfwagen L.T.M 38** would be known now by the German name **Panzerkampfwagen 38 (t)**. Its gun, previously known by the Czech name **3.7 cm Kampfwagenkanone M 38**, would now be designated as the **3.7 cm Kw.K.38(t)**.

18-1

General Description

This general description of the **Panzerkampfwagen 38 (t) Ausf.A bis G und S** is from the manuals D652/32 dated 15 May 1942, D652/38 dated 1 September 1943, and D2015 dated 5 December 1942 - all written near or well after the end of the **Pz.Kpfw.38(t)** production run. Some minor variations between **Ausfuehrung** (such as the radio operator's visor, antenna mounts, and turret originally outfitted for one-man operation) were not specifically identified in these operating manuals.

The **Aufbau** (superstructure) for the **Pz.Kpfw.38 (t), Ausf. A to G** and **S** consists of the upper part of the **Panzerwanne** (armor hull) and the **Turm** (turret). The drive train with engine, transmission and steering unit, and final drives is mounted in/on the lower part of the **Panzerwanne** and, with the suspension, wheels, and tracks, makes up the **Fahrgestell** (chassis).

The Aufbau has space for a crew of four, the **Pz-Fuehrer** (commander/gunner), **Pz-Schuetzen** (M.G.-gunner and loader), **Pz-Fahrer** (driver), and **Pz-Funker** (radio operator). The **Turm** is traversable through 360 degrees and has a **Pz-Fuehrerkuppel** (commander's cupola) for the **Pz-Fuehrer** to observe the terrain.

Armament consists of a **3.7 cm Kw.K. M38(t)** and two heavy air-cooled machineguns **MG 37(t)**. The 3.7 cm gun and a machinegun are mounted in the turret with an elevation arc of -10 to +25 degrees. These weapons can be uncoupled and independently aimed. Ammunition stowage consists of 15 magazines each with 6 rounds of 3.7 cm ammunition, 2700 rounds of 7.92 mm ammunition, for both MG 37(t), and 24 rounds of **Leuchtmunition** (for the signal pistol).

Fahrgestell (Chassis)

The **Panzerwanne** (armor hull) is constructed from **Panzerblechen** (armor plates) of various size and thickness. It is divided into a **Kampfraum** (fighting compartment) and a **Motorraum** (engine compartment) by a **Zwischenwand** (firewall). There are two **Rahmenklappen** (hatches with adjustable vents) in the **Zwischenwand** that allow the **Fahrer** access to the engine from the **Kampfraum**.

The armor plates on top of the rear of the **Panzerwanne** making up the **Motorabdeckung** (engine cover) are removable. Left and right are **Motorabdeckklappe** (engine hatches), each with two hinges that are held by special bolts. Cooling air enters below the overhang of the **Motorabdeckklappe** which protect the engine from rifle or machinegun fire. A screen keeps large foreign debris from entering the engine compartment. The vertical part of both **Motorabdeckklappe** is reinforced on the inside by a 5 mm thick armor plate. **Sperrklappen** (sealing louvers) are installed which can be adjusted by the **Pz-Fahrer**.

Power is provided by a **Praga TNHPS/II** four-stroke, 6-cylinder in-line engine with dry sump lubrication. With a bore of 110 mm and stroke of 136 mm, the swept volume is 7.75 liters with a low compression ratio of 1:6.2. The engine with a single **Solex-Gelaendevergaser** (all-terrain carburetor) was rated at 125 horsepower at 2200 rpm. A hand crank can be used to start the engine from inside the Pz.Kpfw. A mechanical **Drehzahlregler** (governor) limited the engine speed to 2000 rpm. The maximum speed of 42 km/hr is based on an engine speed of 2200 rpm; therefore the top speed at the governed 2000 rpm was only 38 km/hr.

Engine coolant is cooled by a fan in the rear drawing air through a radiator. To maintain the most favorable

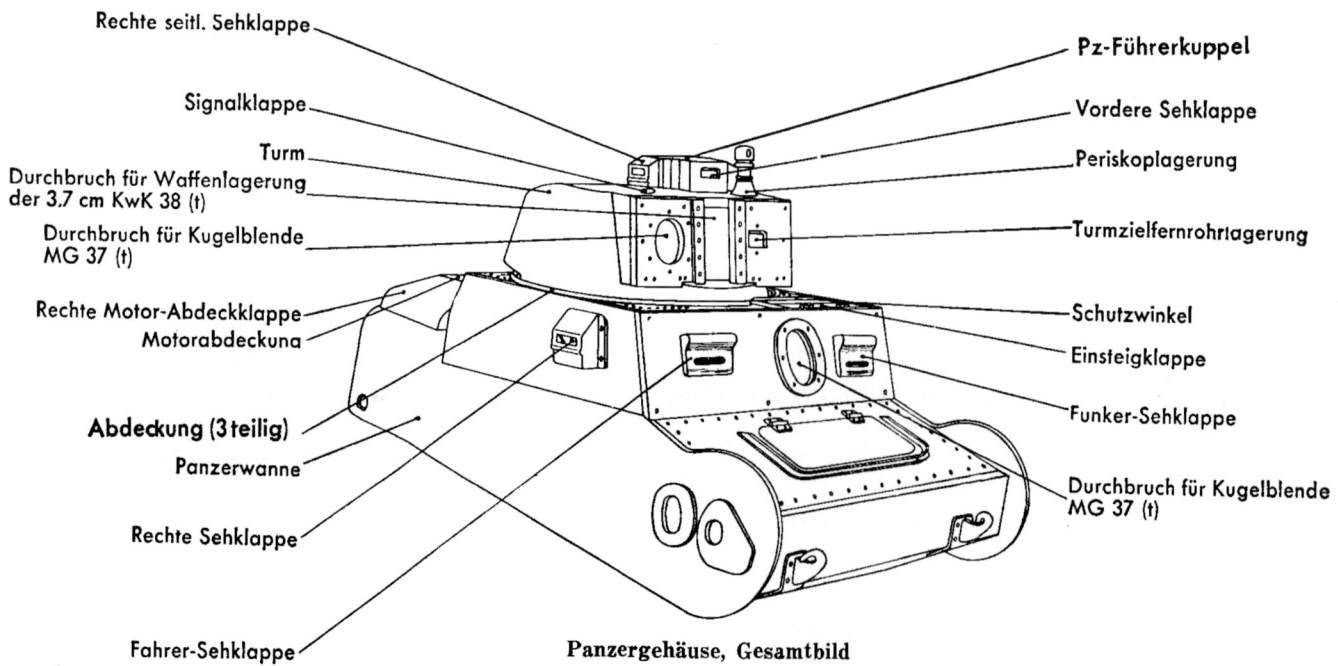

Panzergehäuse, Gesamtbild

operating temperature of 80 to 85 degrees C, the Pz.Kpfw. is outfitted with a device to regulate the coolant temperature from inside the fighting compartment. Older **Ausf.** have internally adjustable louvers, while the newer **Ausf.G** has an adjustable **Schieber** (sliding baffle plate) on the cooling air exhaust.

A **Gelenkwelle** (main drive shaft) runs forward through the fighting compartment to the **Hauptkupplung** (main clutch) on the **Wechselgetriebe** (transmission). The **Praga-Wilson Type CV-TNHP** five-speed transmission has one reverse gear. All the gears are continuously engaged. The I to IV and reverse gear are operated by an associated brake band, the V gear by a slip clutch. The desired gear is selected on the **Vorwaehler** (preselector), and then shifting occurs by pushing down and releasing the clutch foot pedal. A special device inside the transmission prevents engaging two different gears at the same time.

The **Lenkgetriebe** is a clutch-brake steering unit with two steering brake drums and two bypass drive brakes. With a bypass drive brake applied on one side, the **Pz.Kpfw.38(t)** turns in a 9 meter radius curve. For longer radius curves, the bypass drive brake can be intermittently applied. For curves with a tighter radius than 9 meters, the steering clutch is released and a steering brake applied. The **Pz.Kpfw.38(t)** can also be turned on one track by pulling back so hard on the steering lever that the steering brake no longer slips and the final drive no longer turns the drive sprocket and track on one side. There is a second brake band on the steering brake drum that can be independently operated by hand levers or foot brake pedals to stop the Panzer or serve as parking brakes.

Four 775 mm diameter **Laufraeder** (road wheels) with **Gummibandagen** (rubber tires) are mounted on both sides in pairs sharing a leaf spring suspension. The roadwheel discs are made out of 6 mm thick armor plate in order to partially protect the springs and swing arms. The swing arms are designed to continually scrape dirt off the inner side of the **Gummibandagen**. The 535 mm diameter **Leitrad** (idler wheel) is attached with a crank arm to the **Kettenspanner** (track adjuster). Two 220 mm diameter

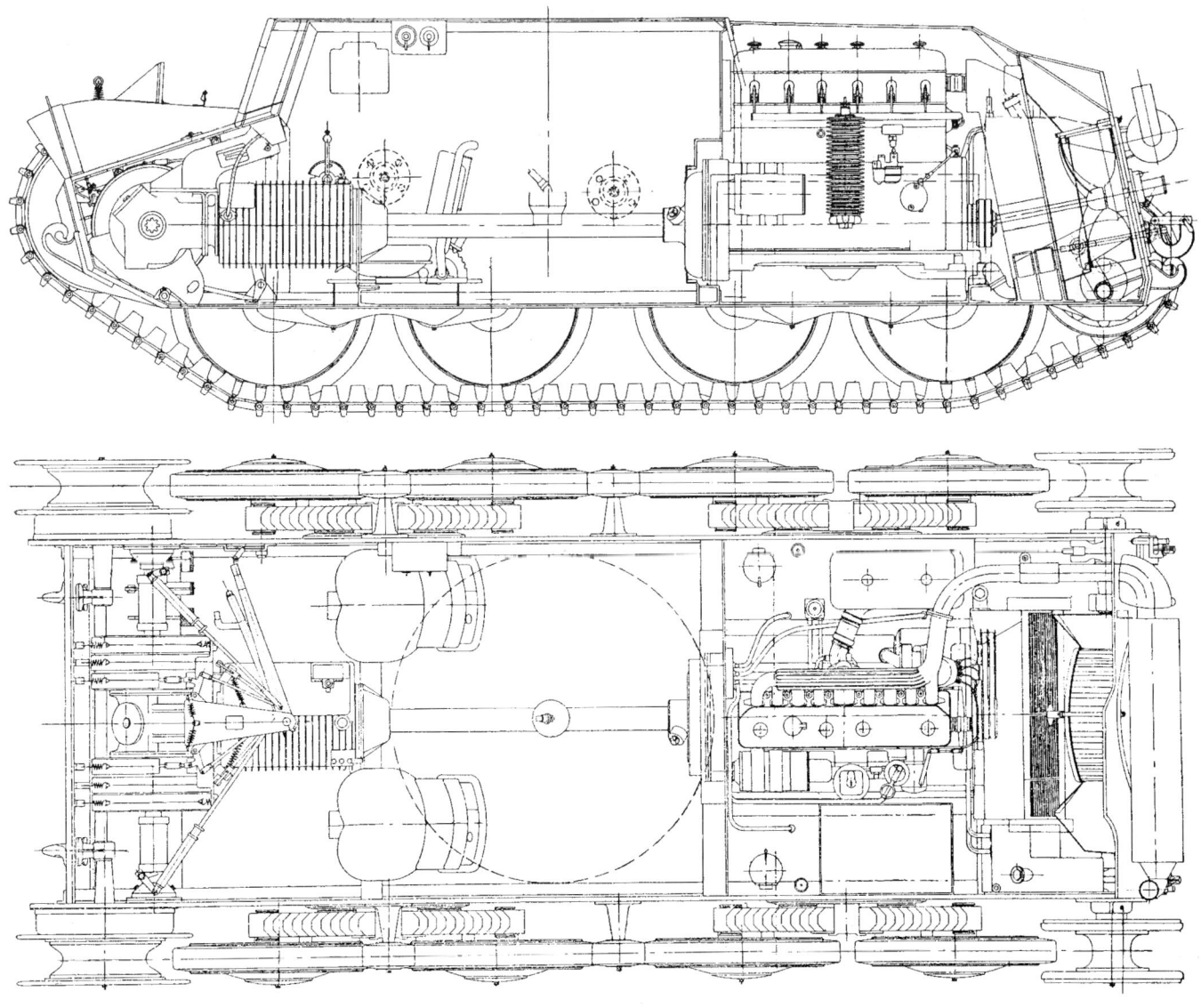

Fahrgestell des Pz. Kpfw. 38 (t)

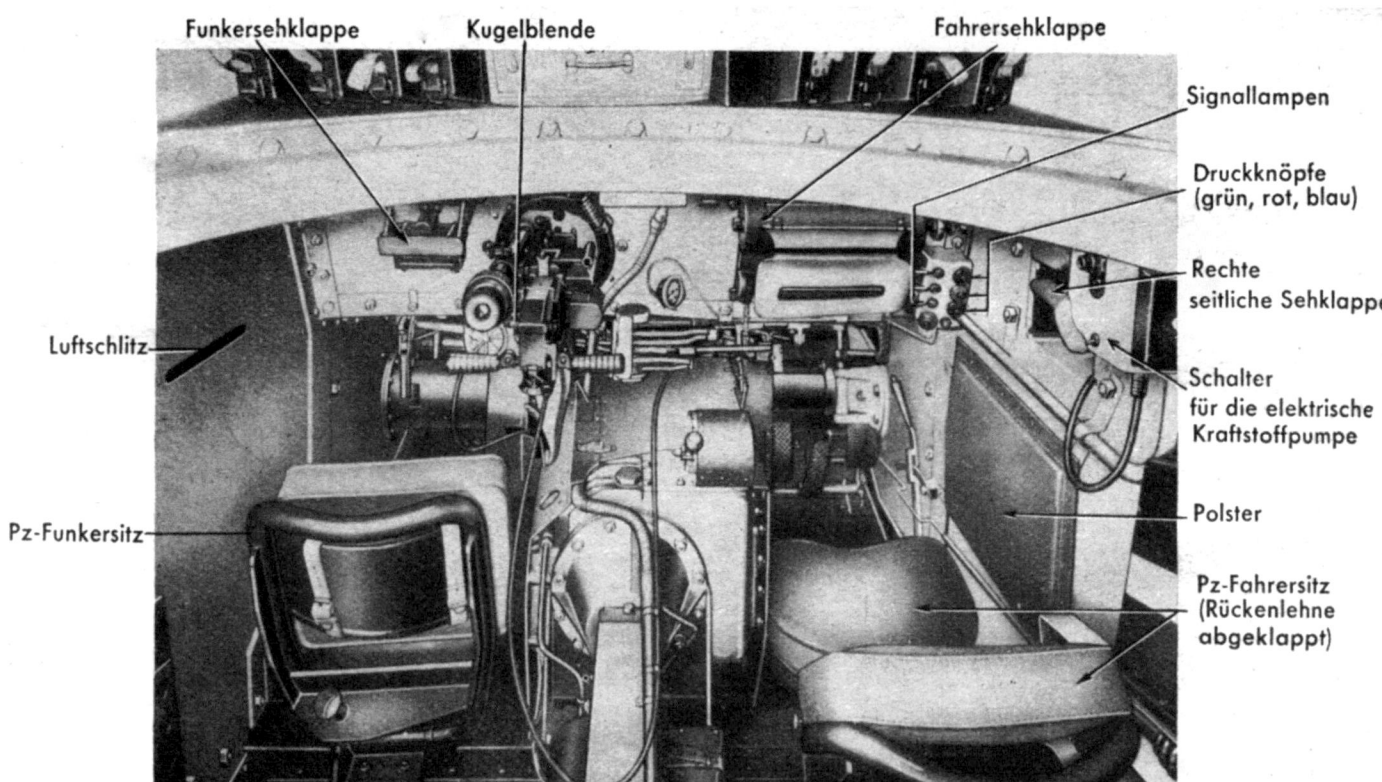

This and Opposite Page: These photographs of a Slovak LT-38 interior were "doctored" by B.M.M. to illustrate a German Army operating manual for the Panzerkampfwagen 38 (t). Having been patterned after the Ausf.D, the Slovak LT-38 was similar but not identical to a Pz.Kpfw.38(t). (NA)

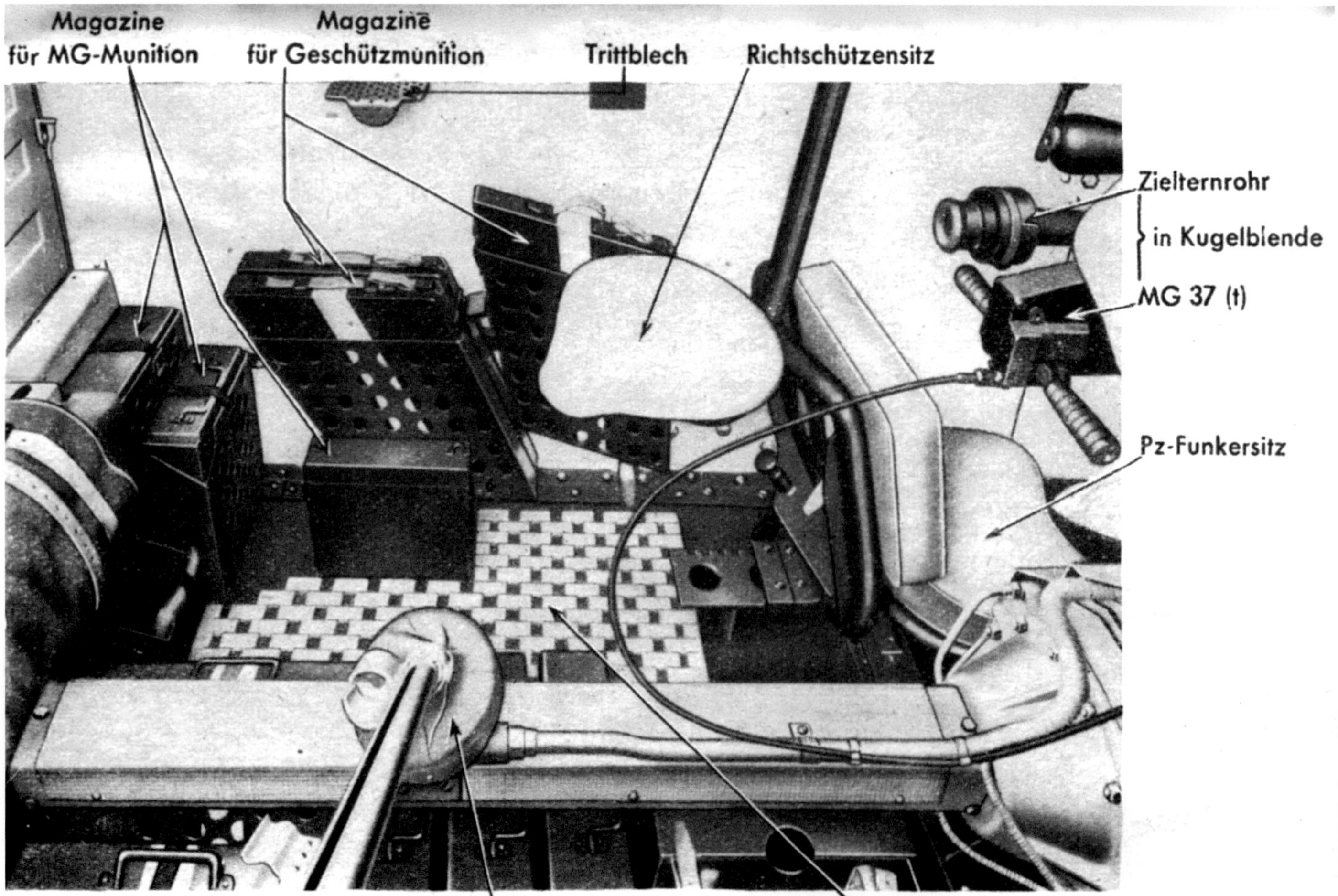

18-4

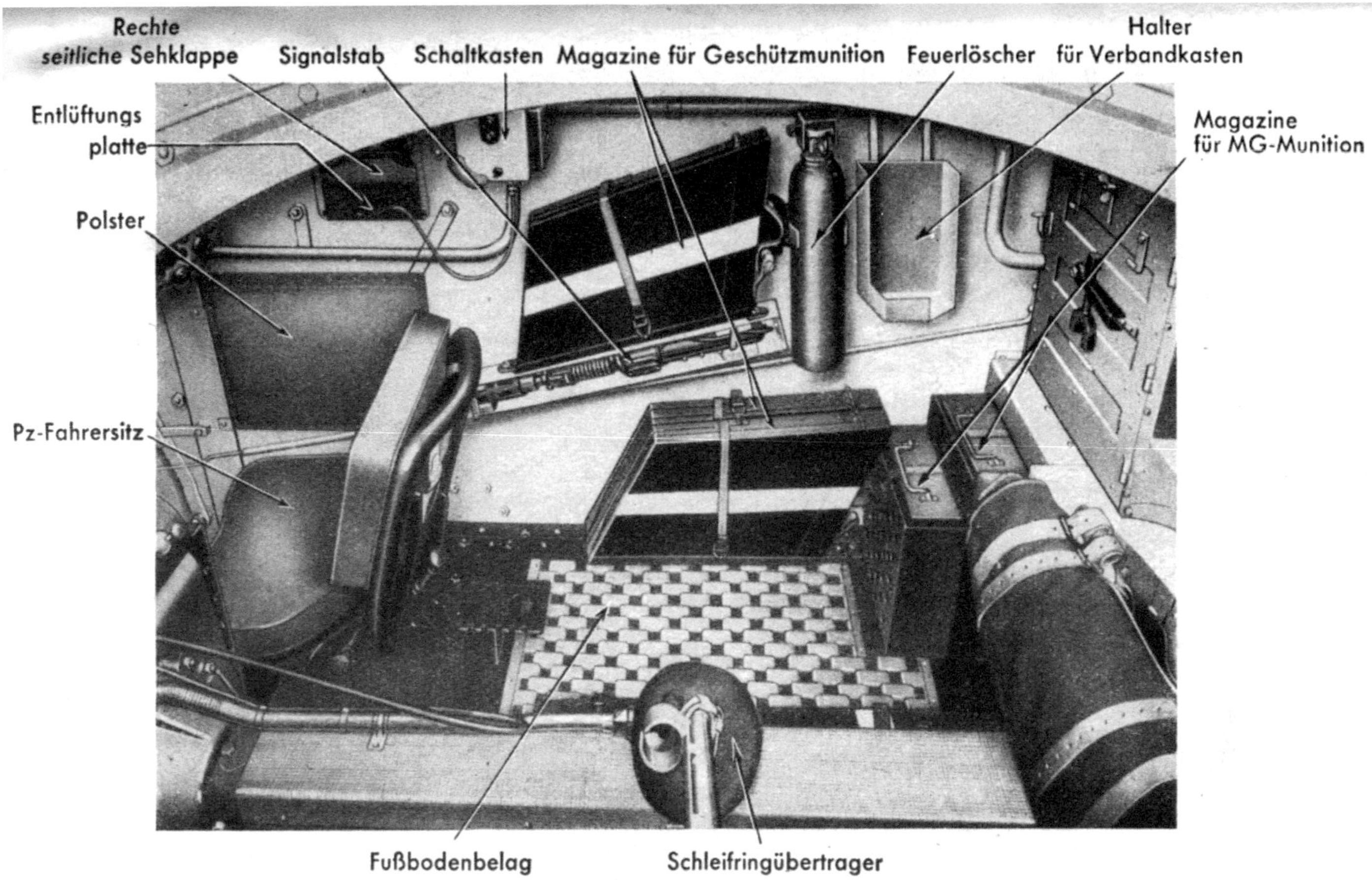

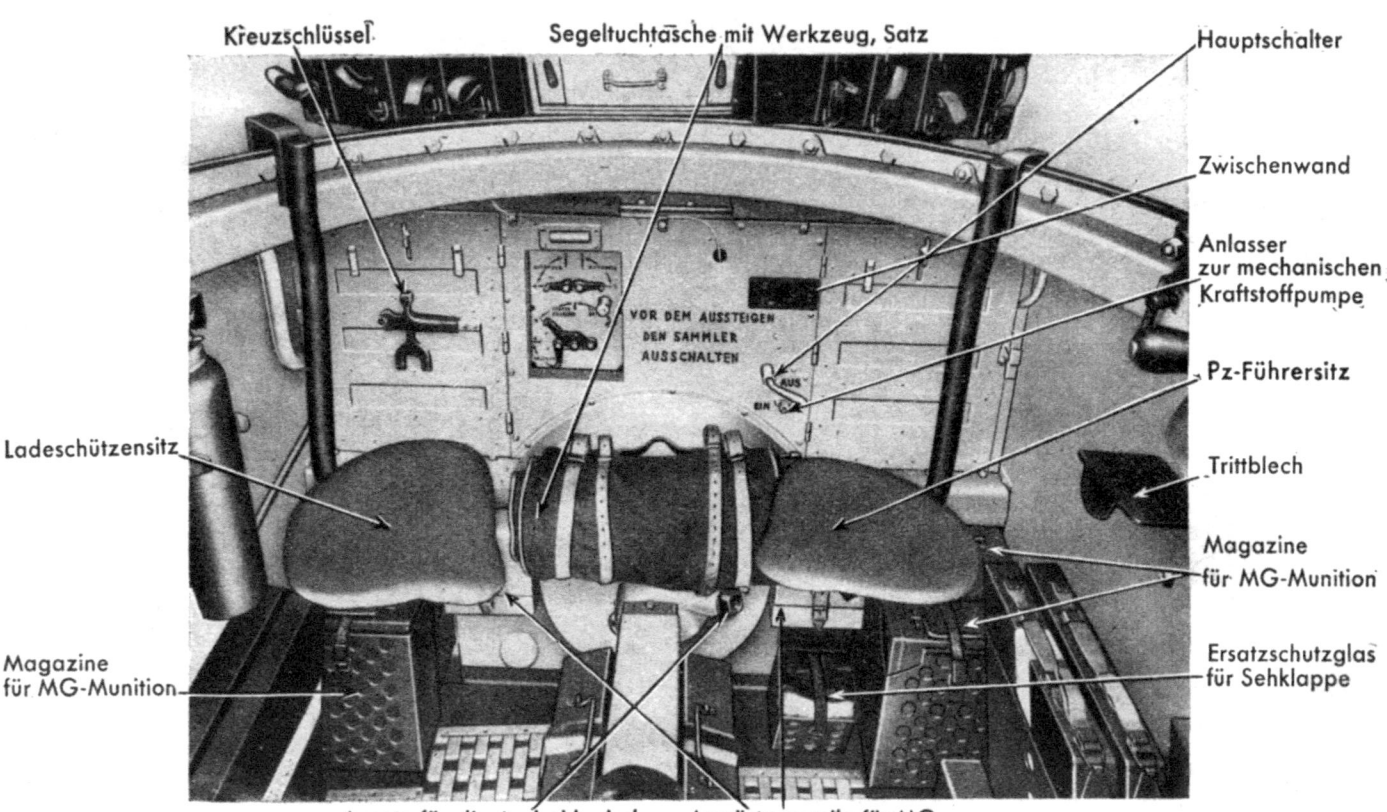

This and Opposite Page: B.M.M. "doctored" these photographs of a Slovak LT-38 interior by drawing in straps to hold the Polster (cushion) to the right of the driver, a Halter fuer Verbandkasten (first aid kit) on the right wall behind the fire extinguisher, and operating instructions on the firewall. Also, details on the left hull side were painted out and an angled rivet backing plate drawn in to fake an Ausf.E to G. (NA)

18-5

Stuetzrollen (return rollers) support the track on its return to the forward drive sprocket. The steel **Kettenglieder** (track links) have a pitch of 104 mm and a width of about 293 mm. They are connected to each other by **Bolzen** (pins) without any heads. A **Sprengring** (circular clip) inside the track link holds the **Bolzen** in place.

The entire length and width of the tracks are covered by **Kettenabdeckungen** (track guards) made of 2 mm thick sheet steel. They are sufficiently braced for the crew members to stand on them. The **Kettenabdeckung** are angled so that rainwater runs off from the inner toward the outer edge.

For driving at night, the **Pz.Kpfw.38(t)** is outfitted with:
o two **Seitenleuchten** (side lights) with spring bases at the front,
o four dismountable **Schlusslichter roter Farbe** (red reflectors) (two at the front and two at the rear) to show the width, and
o a dismountable electrical **Scheinwerfer** (headlight) mounted in the middle of the glacis at the front. During daylight travel, all **Leuchten** and **Schlusslichter** are to be stowed in a felt-lined box on the left side of the engine.

The interior is illuminated by three **Wandleuchten** (wall lights), with one located on the firewall, a second in the turret, and the third above the instrument panel.

At the front and rear there are two **Zughaken** (ram's horn tow hooks), each rated at 5000 kg. Another **Zughaken mit Sicherung** (tow hook with safety catch) is mounted on the rear for towing about 2000 kg.

Kampfraum (Fighting Compartment)

Two vision ports are cut into the plate in front of the **Pz-Fahrer** and **Pz-Funker**. To the right of the **Pz-Fahrer** is another vision port.

A 4 mm wide vision slit is cut into the **Funkersehklappe** (radio operator's vision port) with a stepped aperture in front to deflect lead splash and bullets. A 50 mm **Schutzglas** (protective glass block) protects the eyes against lead splash and fragments.

The **Fahrersehklappe** also has a 4 mm wide vision slit with stepped aperture. Either a 50 mm **Schutzglas**, rotatable **Winkelspiegel** (periscope), or **Schutzscheibe** (tempered glass windshield) can be mounted behind the **Fahrersehklappe**. Two glass mirrors are mounted on a shaft in the top of the **Winkelspiegel** (which can be rotated to replace a damaged mirror) and a stainless steel mirror at the bottom. When the **Fahrersehklappe** is opened for road marches, the large opening is covered by the **Schutzscheibe**.

There is also a **Sehklappe** for the **Pz-Fahrer** mounted on the right side consisting of a armor housing in which a **Doppelspiegel** (double mirror) is mounted. The **Doppelspiegel** is protected by a visor with a vision slit and **Schutzglas**. A **Wischer** (wiper) is installed to clean the mirror. If damaged, the **Doppelspiegel** can be replaced by a spare. There is also a 180 mm by 25 mm vent in the lower part of the armor housing.

The **MG 37(t)**, mounted in a **Kugelblende** (ball mount) in the superstructure front plate, has a traverse of 28 degrees (limited by the belt feed) and an elevation arc of -10 to +10 degrees (limited elevation to prevent hitting the gun tube). A **Kugelzielfernrohr** (telescopic sight) for use by the **Pz-Funker** has 2.6 x magnification, 25 degree field of view, and range graduated in 100 meter increments from 0 to 1700 meters. A firing device with trigger on the left steering lever allows the **Pz.-Fahrer** (driver) to fire the hull **M.G.37(t)** secured to fire at a fixed range adjustable from 300 to 500 meters. The sight for the **Pz-Fahrer** consists of an adjustable ring sight mounted on the front of the glacis plate and a mark in the driver's visor for a back sight.

When complely outfitted with radio equipment, (such as for a company commander), two radio racks could be installed in the **Aufbau**. An upper rack for two receiver sets is suspended from the roof and a lower rack for a sender set is mounted above the guard for the drive shaft. Within the radio equipment circuit is a **Kasten Pz.Nr.10b** (intercom box) and **Fernhoerer-Mikrofon** (headsets and microphones) for the **Pz.-Fahrer**, **Pz.Funker**, and **Pz.-Fuehrer**.

There is a **Signalkasten** (signal box) mounted in the right corner by the **Pz-Fahrer**. It has three push buttons (green, red, blue) and three signal lights (green, red, blue) for communication between the **Pz.Fahrer** and **Pz-Fuehrer**. A **Sprechschlauch** (speaking tube) serves for communication between the **Pz.Schuetzen** and **Pz-Fahrer**.

The **Aufbau** is not gas proof. Four **Gasmasken** are provided for defense against poison gas. A lockable **Einstiegklappe** (entrance hatch) is located in the front left corner of the roof above the **Pz-Funker**. Two adjustable padded seats are provided for the **Pz-Fahrer** and **Pz-Funker**.

Inside on the left sidewall are mounted one holder for three magazines for 3.7 cm ammunition, one step, one box for vision port glass, one holder for the **Leuchtpistole**, one pistol holder by the air slit, and one pistol holder on the upper rear brace.

Inside on the right sidewall are mounted one holder for a first-aid kit, two magazine holders for 3.7 cm ammunition, one holder for fire extinguisher, one holder for signal stick, and one cover guard for electrical components.

Inside on the firewall are mounted one holder for the key, one light, the main power switch, one starter for the mechanical fuel pump, two holders for machinegun parts, and two holders for **MG-Magazine**.

Pulvergase (spent propellent fumes) are adequately cleared by drawing engine cooling air from the fighting compartment, or they can disperse through the **Luftschlitze** (vent slit) on the left sidewall, **Sehklappe** (vision port) on the right sidewall, and out of the turret.

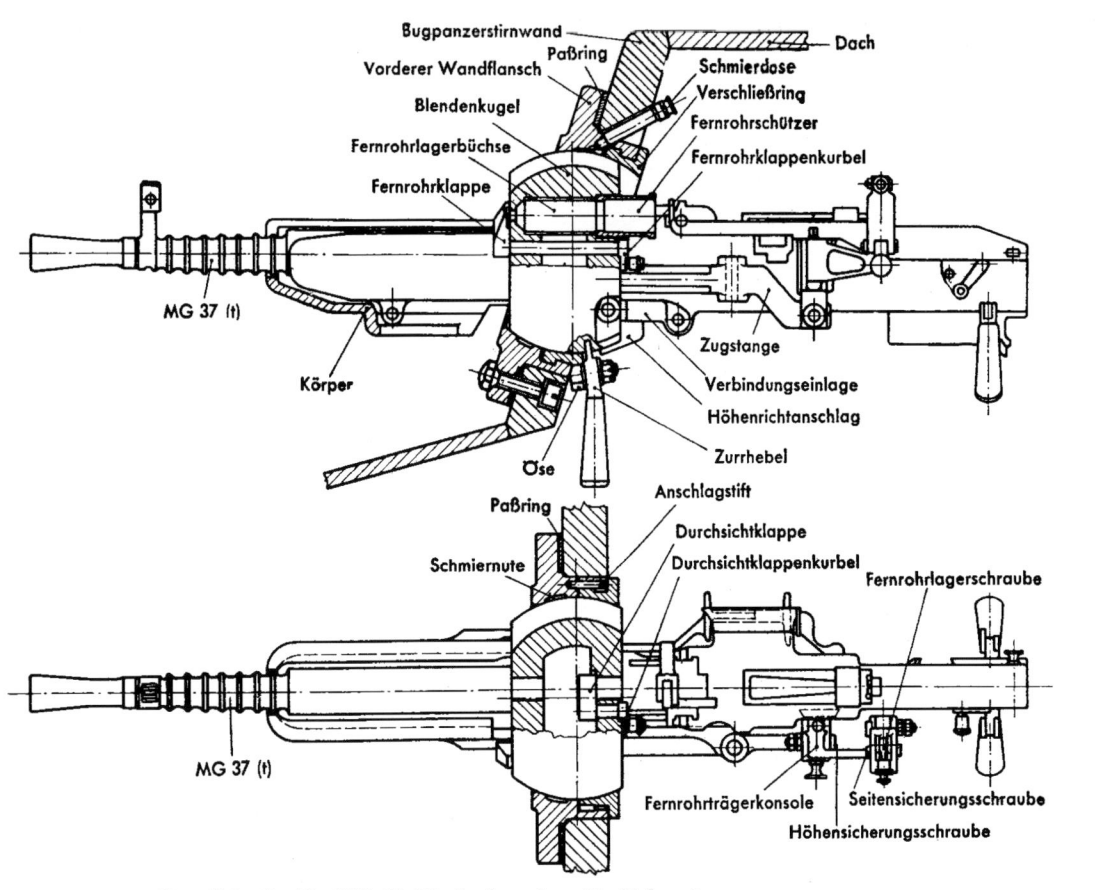

Kugelblende für MG 37 (t) (neben dem Pz-Fahrer)

Turm (Turret)

The **Turmmantel** (armor body) is made out of various size and thickness armor plates that are riveted together. The front plates are set at an angle of 10 degrees and the side plates at an angle of 9 degrees from vertical. A **Kugelblende** (ball mount) for an **MG 37(t)**, a **3.7 cm Kw.K.M38(t)** in a **Geschuetzblende** (gun mantle), and the **Turmzielfernrohrlagerung** (gun sight mount) are all mounted in the turret front. Normally both the machinegun and **3.7 cm Kw.K.** are connected as a pair so that they are elevated together with the geared elevation mechanism turned by a handwheel (35.7 mils per rotation). To quickly engage moving targets, the elevation mechanism can be disconnected for free elevation with the shoulder.

By pulling out the connecting pin, the loader can independently aim and fire the **7.92 mm MG 37(t)** mounted in a **Kugelblende** in the turret. When disconnected from the gun, the **M.G.37(t)** in the **Turmkugelblende** can be traversed about 15 degrees and has an elevation arc of -10 to +25 degrees. An **MG-Zielfernrohr 38(t)** telescopic sight with 2.6 x magnification and 25 degree field of view is graduated from 200 to 1500 meters in 100 meter increments. All gun sights had an inverted 'V' as a central aiming mark.

The **Turm** is mounted on the **Turmkugellager** (ball-bearing race) for 360 degree traverse. The inner ring of the **Turmkugellager** (inner diameter of 1210 mm) has 405 Module 3 teeth for turning the turret with the traversing gear. The **Turmschwenkwerk** (traversing gear) is mounted on the side of the turret. An **Entkuppler** (decoupler) in the **Turmschwenkwerk** is used for free turret traverse by pushing on the **3.7 cm Kw.K.M38(t)**. When the gun is fired, a **Kugelbahnbremse** (ball race brake) is used to hold the tur-

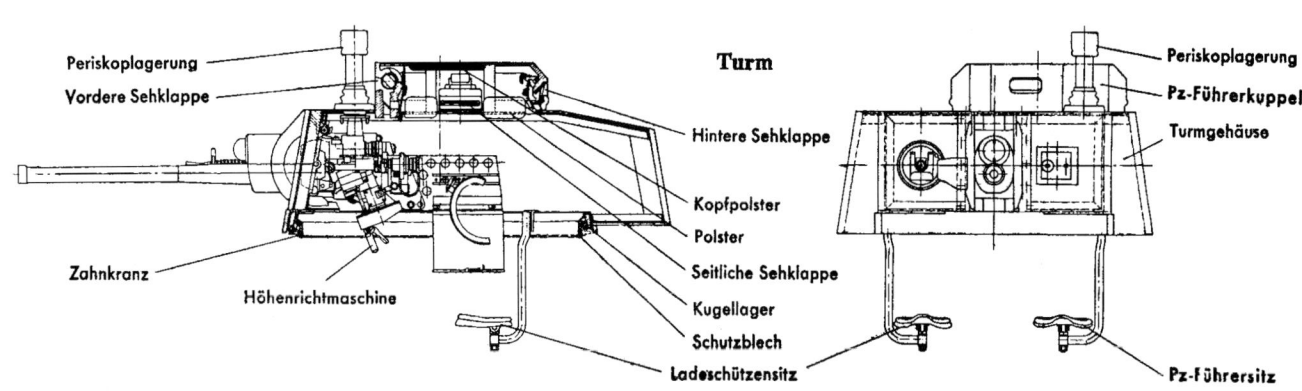

18-7

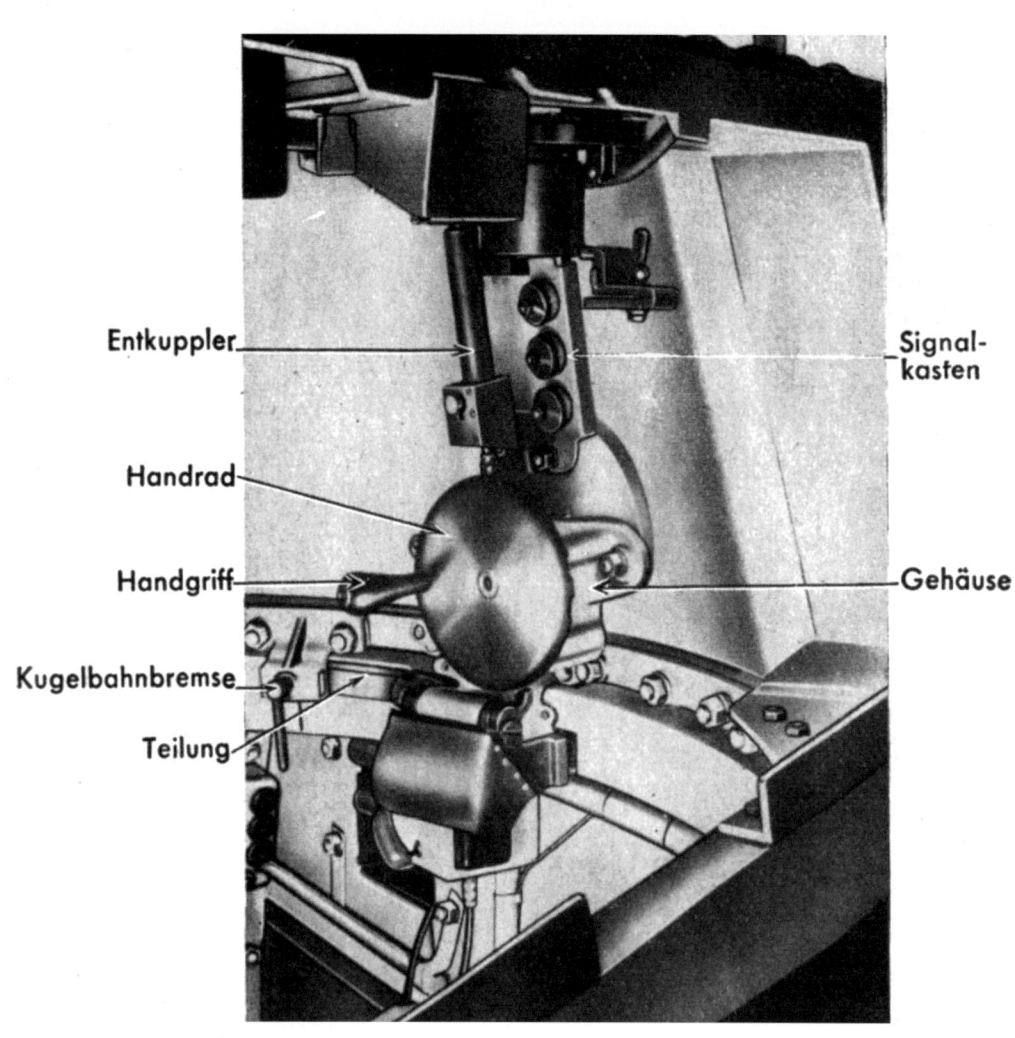

This and Opposite Page: These photographs of a Slovak LT-38 interior were also used by B.M.M. to illustrate a German Army operating manual for the Panzerkampfwagen 38 (t). This explains why there isn't any German radio equipment inside this "faked" Pz.Kpfw.38(t). The Signalkasten (signal box) with green, red, and blue lights, used for communication between the commander and driver, was more reliable than an intercom system, which was vulnerable to being disabled by the shock of near misses from artillery fire. (NA)

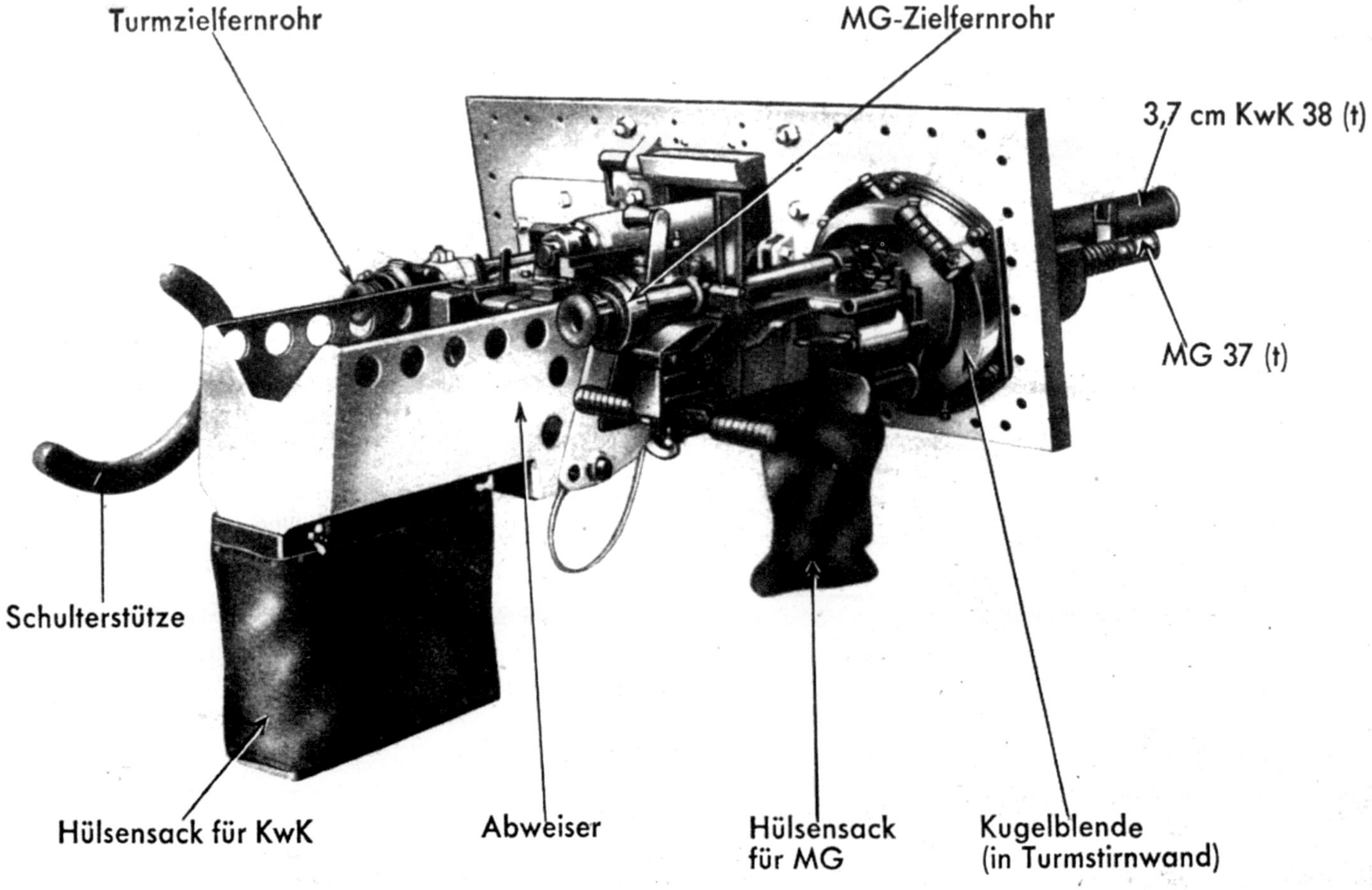

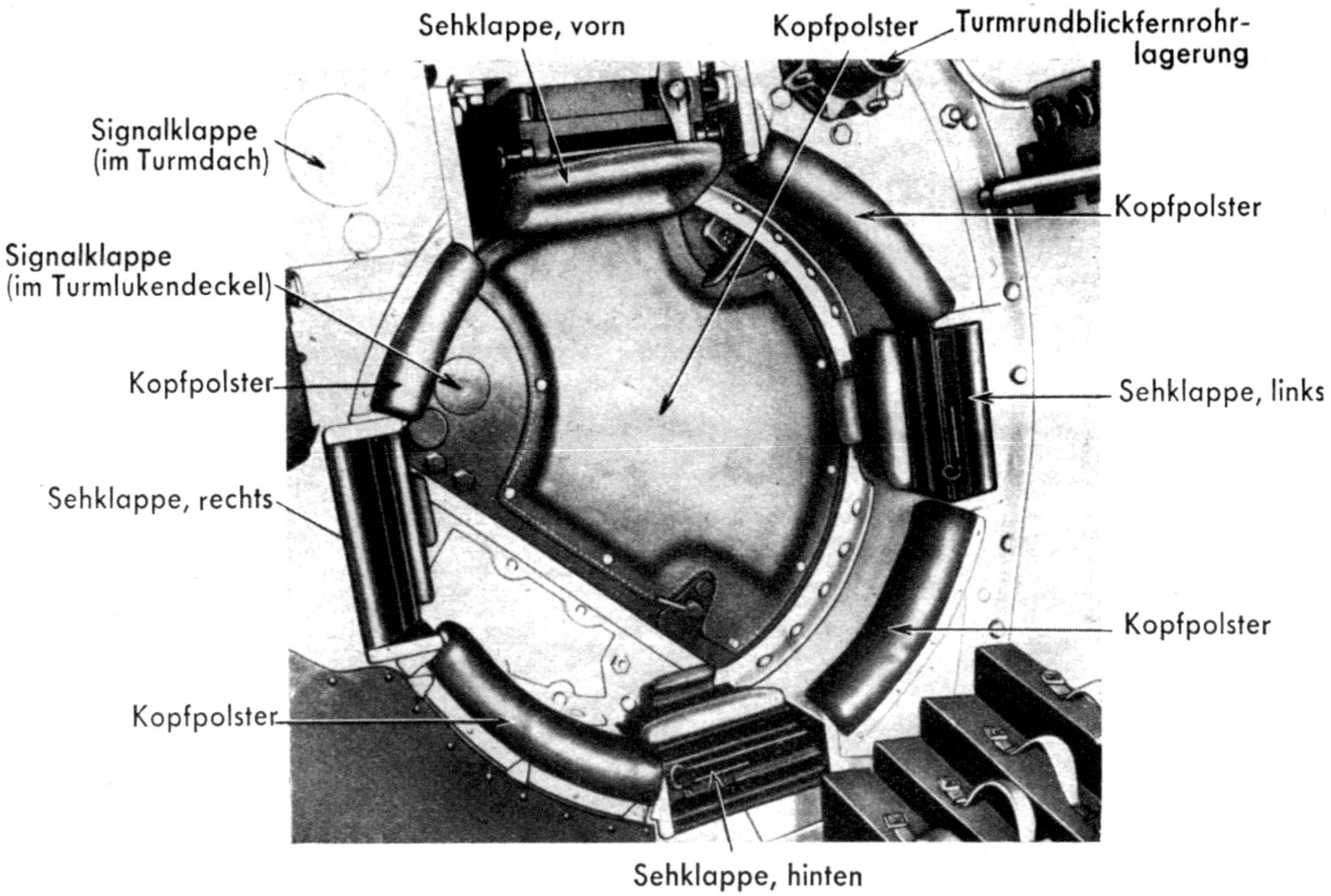

Signalklappe (im Turmdach)

Sehklappe, vorn

Kopfpolster

Turmrundblickfernrohrlagerung

Signalklappe (im Turmlukendeckel)

Kopfpolster

Kopfpolster

Sehklappe, links

Sehklappe, rechts

Kopfpolster

Kopfpolster

Sehklappe, hinten

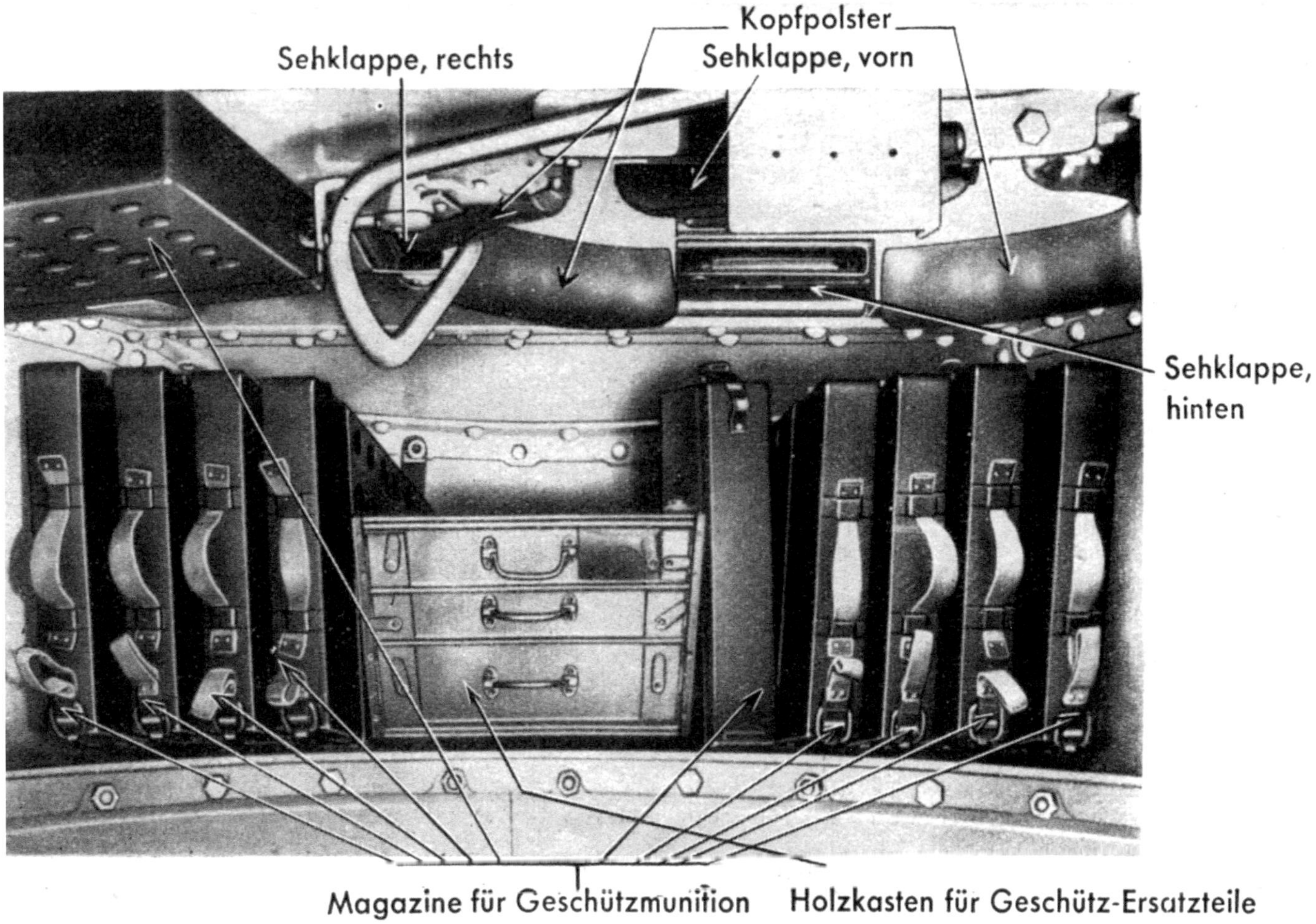

Sehklappe, rechts

Kopfpolster
Sehklappe, vorn

Sehklappe, hinten

Magazine für Geschützmunition

Holzkasten für Geschütz-Ersatzteile

18-9

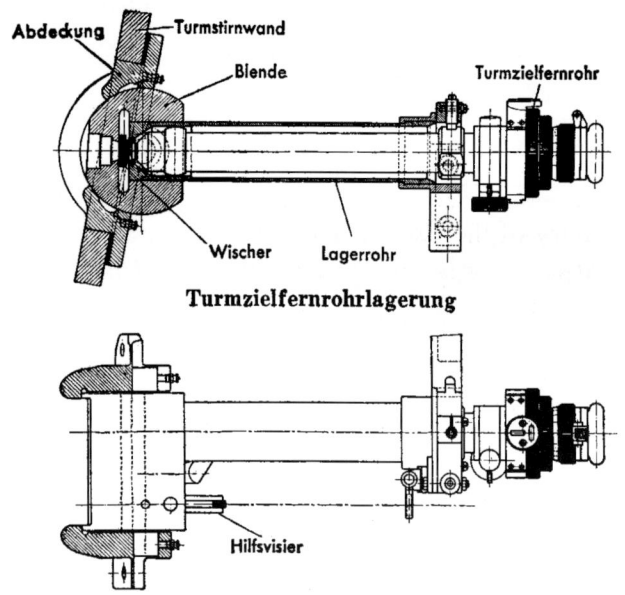

ret in place in any direction desired. A **Sperrvorrichtung** (travel lock) completely blocks the turret from traversing during movement and is only to be disengaged when combat ready.

The **Turmzielfernrohr 38(t)** telescopic sight with 2.6 x magnification and 25 degree field of view is graduated from 200 to 2000 meters in 100 meter increments for the 3.7 cm gun and 200 to 1500 meters for the coaxial machinegun. If the **Turmzielfernrohr** is damaged, a plug can be removed through which the gun can be fired at ranges up to 1000 meters using a **Hilfsvisier** (auxiliary open sight). The gun can be fired at a rate of up to 15 rounds per minute.

A **Pz-Fuehrerkuppel** (commander's cupola) with four opposing **Sehklappen** (vision ports) is mounted on the turret roof. The front **Sehklappe** consists of an armor housing in which a rotatable **Winkelspiegel** is mounted. Two glass mirrors are mounted on a shaft in the top of the **Winkelspiegel** (which can be rotated to replace a damaged mirror), and there is a stainless steel mirror at the bottom. The **Sehklappen** on the left, right, and rear of the **Pz-Fueherkuppel** consists of an armor housing with two fixed mirrors (to create a periscope) and a cover with a vision slit.

The **Turmlukendeckel** (hatch) in the top of the cupola is hinged at the rear. One of the hinges is designed to hold the hatch lid open in three different positions. There is a 70 mm diameter **Signalklappe** (signal port) in the cupola hatch lid. It can be used as a vent when firing or for signaling with signal flags or firing the **Leuchtpistole**.

A 105 mm (or 100 mm) diameter **Signalklappe** (signal port) is located in the turret roof. It can be opened for signaling with a **Signalleuchte** (signal light) or for ventilation to reduce the carbon monoxide built up from firing.

A **Turmrundblickfernrohr 38(t)** (pivoting traversable periscope) is mounted inside an armor housing on the turret roof. This panoramic periscope with 2.6 x magnification and 25 degree field of view can be rotated

for observation in any direction. Starting with the **Ausf. B**, a **Panzerkappe** (armor cap) was added to protect the periscope head (which on an **Ausf.A** protruded out the top of the armor housing). The **Panzerkappe** has two rows of ball bearings to aid in freely traversing with the panoramic periscope.

Two padded seats, one for the **Ladeschuetze** (loader) and one for the **Pz-Fuehrer** (commander), are mounted on tubular supports that are suspended from the turret ring. There is a **Signalkasten** (signal box) mounted to the left of the **Pz-Fuehrer**. It has three push buttons (green, red, blue) and three signal lights (green, red, blue) for communication between the **Pz.-Fahrer** and **Pz.-Fuehrer**.

Gun ammunition, tools, and spare parts are stowed in the rear of the turret. Six 3.7 cm rounds are carried in a "ready-rack" below the turret roof above the coaxial machinegun.

The gap between the turret and the roof of the hull is protected by a three-piece **Abdeckung** (turret ring guard - introduced during the **Ausf.C** production run) which prevents rifle and machinegun fire from getting through the gap to the **Turmkugellager** (ball bearing race).

Armor Protection

The basis for the thickness and type of armor on the **Pz.Kpfw.38(t)** was all-round protection (15 mm on the sides and rear) against penetration by 7.92 mm AP bullets and frontal protection (25 mm face-hardened) against 2 cm class anti-tank weapons. Therefore it came as no surprise to the designers that the Polish 3.7 cm anti-tank guns could penetrate the front and side armor. However, the crews manning the Panzers had other ideas about adequate protection and following the campaign in Poland had requested

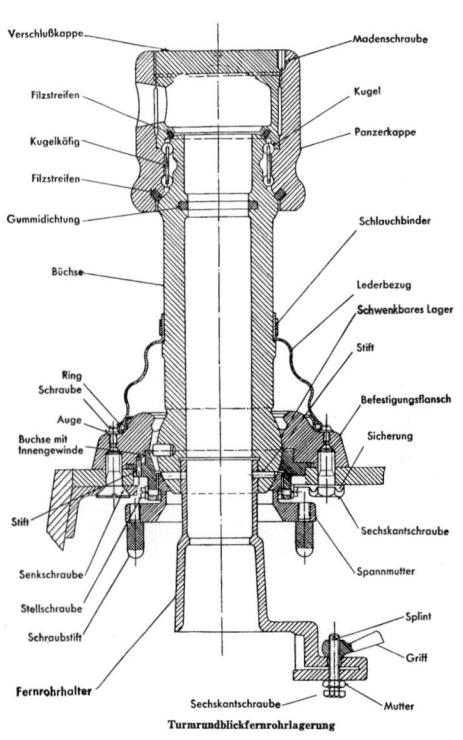

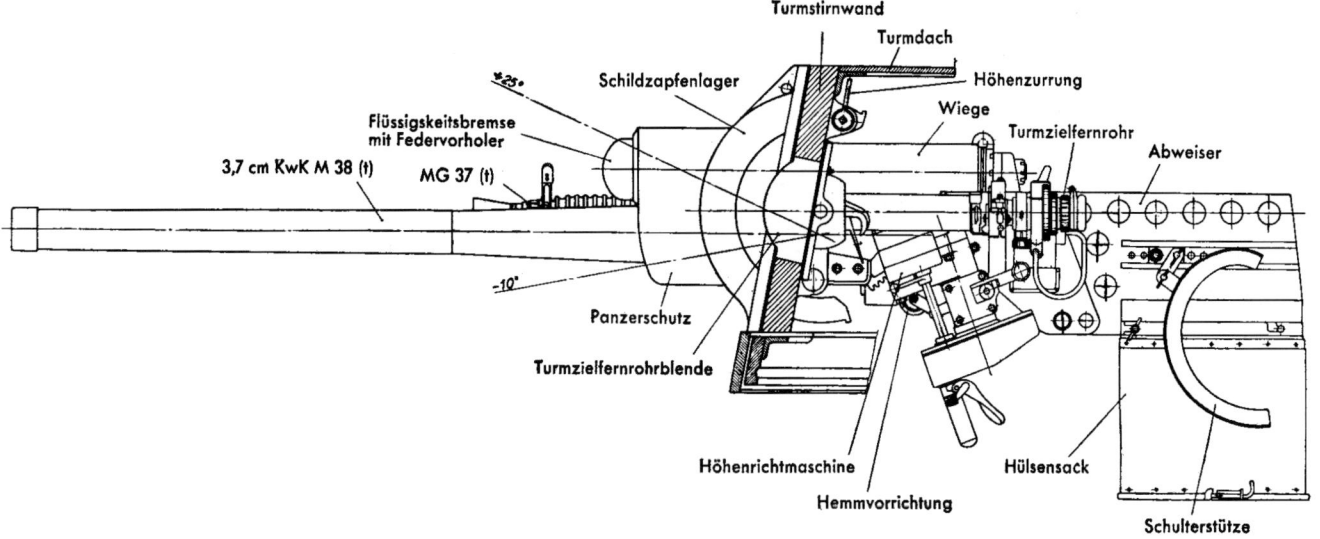

3,7 cm KwK M 38 (t) und MG 37 (t)

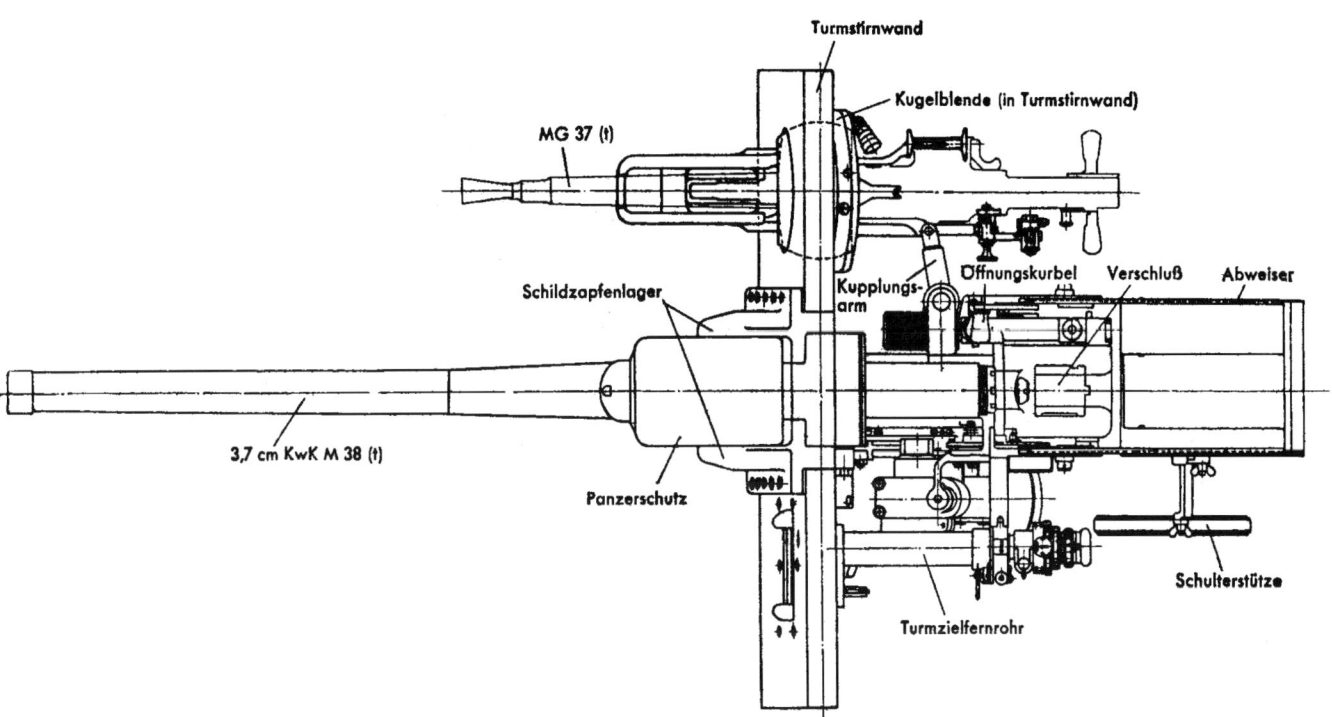

increased armor protection.

The ability of the Waffenamt to meet this request was addressed in the following excerpt from a report on the **Erfahrung Polnische Feldzug** (Experience in the Polish Campaign) dated 25 October 1939: *Backfitted armor reinforcement is only possible in an expedient form on the hull front but not the turret front. By changing the design of future models, it will definitely be possible to increase the armor on the hull front, but because of weight considerations it is still not clarified if an increase is possible for the driver's front plate. However, it is impossible to increase the armor on the turret front because it would make the Panzer nose heavy, impair cross-country mobility, and reduce automotive reliability. Completion of the necessary design changes for increased armor protection won't be accomplished until 1941.*

Expedient reinforcement of the hull front was introduced with the **Pz.Kpfw.38(t) Ausf.C** and continued with all of the **Ausf.D** in 1940. Just like on the **Pz.Kpfw.III Ausf. H** and **Pz.Kpfw.IV Ausf.D** and **Ausf.E**, one armor supplier may have introduced the change before the other. There were two different armor suppliers for the **Pz.Kpfw.38(t)**: Poldihuette, A.G. in Kladno and Witkowitzer Bergbau near Ostrau.

An interim solution was also implemented with the **Pz.Kpfw.38(t) Ausf.E, F,** and **S** with 25 mm face-hardened plates added to the 25 mm thick base plates on the turret front plates, straight driver's front plate, and hull front plate. In addition, the thickness of the turret sides and rear was increased and a 15 mm plate added to the superstructure sides on the **Pz.Kpfw.38(t) Ausf.E** and **F**.

The final design changes were implemented into the

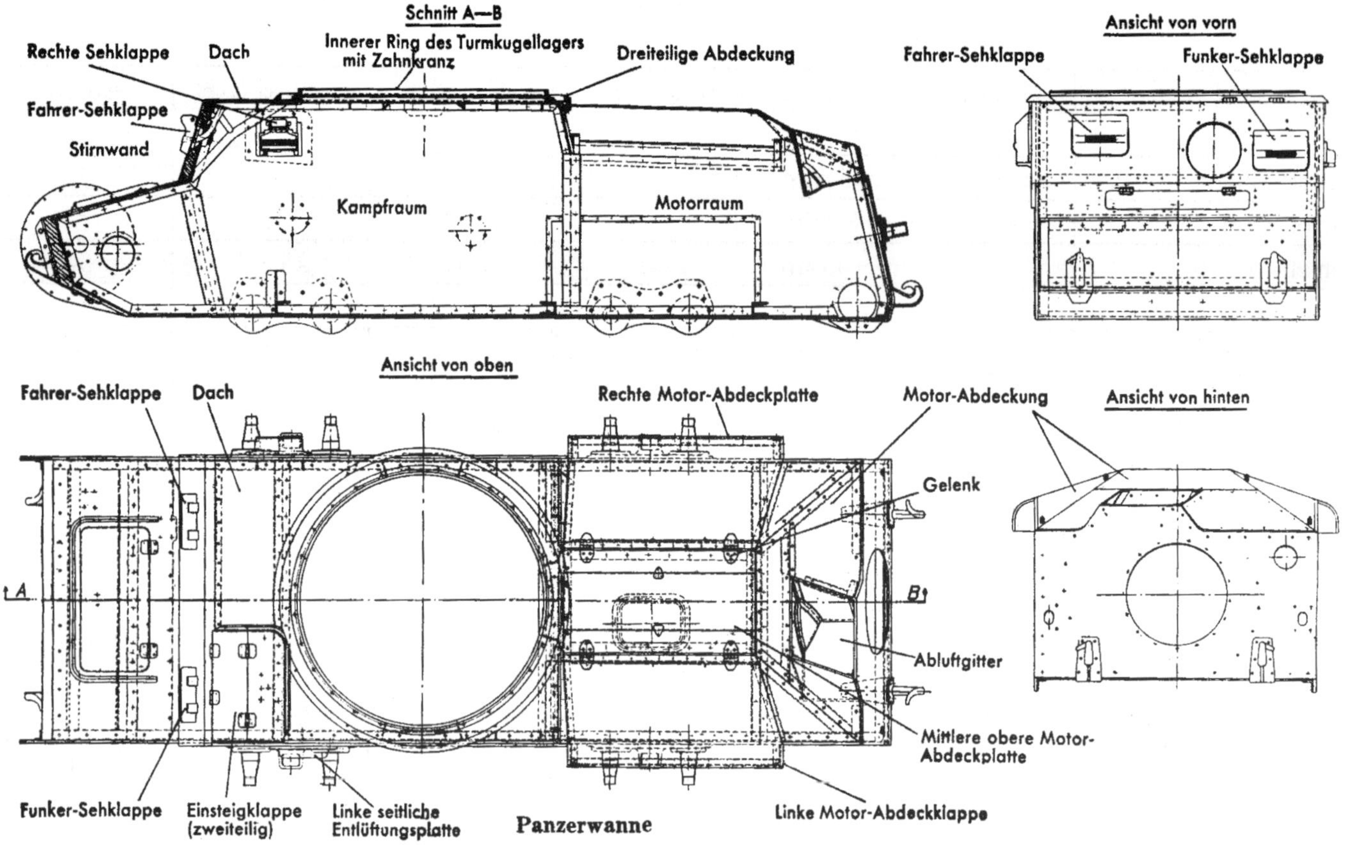

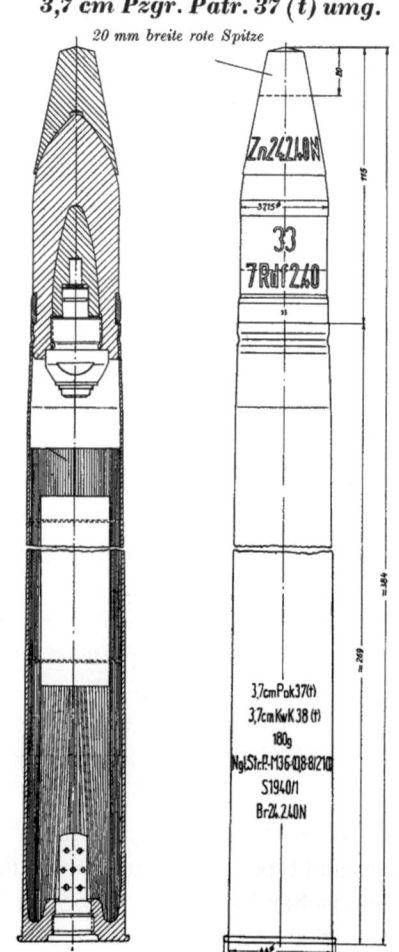

Pz.Kpfw.38(t) Ausf.G ordered in 1941. Homogeneous 50 mm thick plates replaced the face-hardened 25 mm plus 25 mm plates on the turret, superstructure, and hull front. The turret side plates on the **Ausf.G** were 30 mm thick and turret rear 25 mm thick. Additional 15 mm plates were still bolted to the superstructure sides, but the rest of the hull sides and rear were still only 15 mm thick - just like the **Ausf.A**.

3.7 cm Ammunition

Originally the **3.7 cm Kw.K.38(t)** fired a **3.7 cm Pzgr.Patr.37(t)** (weighing 0.850 kg with a 8 gram high-explosive filler at a muzzle velocity of 741 mps) capable of penetrating 28 mm armor plate set at an angle of 30 degrees at a range of 600 meters. Excessive smoke from the muzzle of the gun obscured observation after firing, and there wasn't any tracer on this APHE projectile. As revealed in the manual D430/8 dated 15Feb42, the original **Pzgr.** projectile was modified to add a **Lichtspurhuelse Nr.1** (tracer), enlarge the chamber for a 13 gram high-explosive charge, and add a ballistic/penetrating cap to reduce wind resistance and improve penetration of face-hardened armor plates. This improved **3.7 cm Pzgr.Patr.37(t) umg.** (weighing 0.815 kg) fired from **3.7 cm Kw.K.38(t)** with a muzzle velocity of 750 m/s was capable of penetrating (at 30 degrees) 41 mm at a range of 100 meters, 33 mm at 500 meters, and 27 mm at 1000 meters.

A **3.7 cm Sprgr.** (high explosive shell) had also been developed and produced for the Czech 3.7 cm tank and anti-tank guns. On 16 December 1939, 8.Panzer-Division requested: *Pz.Abt.67 outfitted with the* **tschech. 3.7**

Plate Location	Angle	Ausf.A & B	Ausf.C & D	Ausf.S	Ausf.E & F	Ausf.G
Cupola Top	90	8 mm	=	=	=	12?
Cupola Sides	0	15 mm	=	=	=	=
Turret Roof	90 + 80	8 mm	=	=	12	=
Turret Front	10	25 mm	=	25+25	25+25	50
Turret Sides	9	15 mm	=	=	30	=
Turret Rear	9	15 mm	=	=	25	=
Superstructure Roof	90	8 mm	=	=	=	=
Driver's Front	17.6	25 mm	=	25+25	25+25	50
Glacis	74	12 mm	=	=	=	=
Hull Front	15	25 mm	40	25+25	25+25	50
Lower Hull Front	66	15 mm	=	=	=	30?
Superstructure Side	0	15 mm	=	=	15+15	15+15
Engine Hatch	0-72	10 mm	=	=	=	12?
Hull Side	0	15 mm	=	=	=	15
Wheel Discs		6 mm	=	=	8	8
Engine Deck	8	8 mm	=	=	=	10
Upper Hull Rear	61	10 mm	=	=	=	=
Hull Rear	14	15 mm	=	=	=	=
Belly	90	8 mm	=	=	=	=

cm Kw.K.Mod.38 has 16,000 rounds of **3.7 cm Pzgr.Patr. Mod.37**. Based on experience in the **Polen Feldzug** the issue of **Sprgr.Patr.** is urgently needed. It is therefore requested that 4000 rounds of **tschech. 3.7 cm Sprgr.Patr. f.Kw.K.** be issued in exchange for 4000 rounds **3.7 cm Pzgr. Patr.Mod.37**. OKH, Gen.St.d.H., Generalquartiermeister responded on 20 December 1939: *The available stockpile of **3.7 cm Pzgr.Patr.M37(t)** is very small. It should be used sparingly and solely against armored targets. A larger stockpile of **3.7 cm Spgr.Patr.M34(t)** is available. As needed, **3.7 cm Sprgr.** can be requisitioned from Gen.Qu. Exchange of the 4000 rounds should be conducted through H.Ma.Wulfen.*

By March 1941, a **3.7 cm Pzgr.40/37(t)** weighing 0.368 kg was developed and produced for the **3.7 cm Kw.K.38(t)**. These projectiles had a subcaliber tungsten-carbide core which disintegrated during penetration. Due to its low mass and high muzzle velocity of 1020 mps, its penetration ability dropped off rapidly with range; 64 mm armor plate (at 30 degrees) at a range of 100 meters and 37 mm at 400 meters. A table released in the Summer of 1941 advising the troops of the effective ranges that various anti-tank weapons and types of ammunition could penetrate the **schwerer russischer Panzerkampfwagen T 34** listed the **3.7 cm Kw.K.38(t)** with **Pzgr.40/37(t)** as being capable of penetrating the hull side (but no other location) at ranges up to 300 meters.

Production

CKD was busy assembling the first series of 150 TNH-P light tanks that had been ordered for the Czech army when Germany occupied Bohemia and Moravia on 15 March 1939. Needing additional gun-armed light tanks, the Heeres Waffenamt awarded a contract to B.M.M. (Boehmisch-Maehrische Maschinenfabrik - the German wartime designation for CKD) to complete production of the first series of 150 **Panzerkampfwagen L.T.M.38.**

In a note to the head of the **Waffenamt** on 17 January 1941, **Wa Stab Ia** reported on the expected decrease in output of **3.7 cm Kw.K.38(t)** guns in January and February 1941, and thereby a record was preserved revealing when contract extensions had been awarded for **Pz.Kpfw.38(t)** production:

*Skoda has been awarded the following contracts for the **3.7 cm Kw.K.38(t)**:*
1. 325 by Kolben-Danek (contract dated 12Jun39)
2. 275 by direct contract (contract dated 6Oct39)
3. 250 as a contract extension.

*This third contract from the **Waffenamt** was awarded on 25 July 1940 (but first sanctioned by **In 6** on*

Production Series				
Serie	Ausf	No.	Fgst.Nr.	Produced
1.	A	150	1-150	May-Nov39
2.	B	110	151-260	Jan-May40
3.	C	110	261-370	May-Aug40
4.	D	105	371-475	Sep-Nov40
5.	E	275	476-750	Nov40-May41
6.	F	250	751-1000	May-Oct41
	S	90	1001-1090	May-Sep41
7.	G	269	1101-1359	Oct41-Mar42
7.	G	47	1480-1526	May-Jun42

27 November 1940, after it was made known that further *Pz.Kpfw.38(t)* should be produced). In the interim Skoda was using their forge to complete Flak guns that needed to be completed first. Therefore delays occurred in starting work on the contract extension, further exasperated by delays in receipt of control numbers.

Decreased production in January and February is not a problem, because 150 *3.7 cm Kw.K.38(t)* are stored in the **Heereszeugamt Magdeburg**. This stock is still available, even though in the interim 90 *Pz.Kpfw.38(t)* for Sweden and 11 (actually 10) *Pz.Kpfw.38(t)* for Slovakia have been confiscated and outfitted with *3.7 cm Kw.K.38(t)* guns. Both countries had placed orders for Panzers without guns.

Another 1000 **Pz.Kpfw.38(t) Ausf.G** were ordered from B.M.M. in 1941: 500 in the **7.Serie** under contract 210-3951/41 H and 500 **Pz.Kpfw.38(t) Ausf.H** in the **8.Serie** under contract 210-3952/41 H.

On 12 April 1939, CKD reported the status of TNH-P Series production: *We have a total of 1157 drawings. Both our shop and Slany are finishing all sub-assemblies. 38 chassis are being assembled. 11 vehicles are already running, but are not completed. Supplies of armor plates are coming in better but still not in sufficient quantity. For example, Poldihuette should have shipped 90 hulls, but sent only 43, and 80 complete turrets, but sent only 61. Witkowitzer sent only 18 hulls instead of 40 and 9 turrets instead of 40.*

The first nine **L.T.M.38** were accepted by the **Heeres Waffenamt** inspectors in May 1939, followed by 12 in June, 39 in July, 18 in August, 31 in September, 30 in October, and the last 11 of the **1.Serie** of 150 **L.T.M.38 Protektorat** in November 1939.

As already announced in October 1939, plans were made to complete the first 15 of the **2.Serie** in January 1940, with the rest following at the rate of 25 per month. However, only 10 **L.T.M.38 Protektorat** were accepted in January 1940 because of the difficulty Skoda had in delivering **3.7 cm Kw.K.** guns to Kolben-Danek on time. On 25 April 1940, **Wa J Rue (WuG 6)** reported that out of those authorized by orders from AHA/AgK/In6, 535 **Pz.Kpfw.38(t)** still needed to be completed and delivered after 1 April 1940. As shown in the following table the seven production series **Pz.Kpfw.38(t)** plus 90 confiscated from the Swedish order were completed from May 1939 to June 1942.

Production Statistics				
	1939	1940	1941	1942
Jan		10	45	60
Feb		24	50	60
Mar		31	53	29
Apr		30	49	0
May	9	30	78	21
Jun	12	30	65	26
Jul	39	30	65	0
Aug	18	35	64	
Sep	31	35	76	
Oct	30	44	53	
Nov	11	27	50	
Dec	0	44	50	
Total	150	370	698	196

In February, March, and April 1941, **Wa J Rue (WuG6)** reported that difficulties had occurred in starting production of the 90 confiscated "**Schweden-Fahrzeugen**". These **Pz.Kpfw.38(t) Ausf.S** were then completed in parallel with **Pz.Kpfw.38(t) Ausf.F** from May to September 1941. **Waffenamt** inspectors also included some of the 30 LT vz. 38 tanks ordered by the Slovak Free State in their monthly acceptance statistics for **Pz.Kpfw.38(t)** production from October 1940 to June 1941.

Seventeen **Pz.Kpfw.38(t) Ausf.G Fgst.Nr.1331** to **1347** were accepted by 4 March 1942, one **Pz.Kpfw.38(t) Ausf.G Fgst.Nr.1348** was sent to Alkett for trials on 9 March 1942, and eleven **Pz.Kpfw.38(t) Ausf.G Fgst.Nr.1349** to **1359** were accepted by 11 March 1942. **Pz.Kpfw.38(t)** production was then interrupted in March through May to complete a series of 120 **Pz.Sfl.2** but resumed again in May and June 1942 with 47 additional **Pz.Kpfw.38(t) Ausf.G** being completed before production was entirely shifted to **Pz.Sfl.2**. production. (Refer to Panzer Tracts 7-2 for additional information on the **Pz.Sfl.2**.)

<u>Modifications During Production Run</u>

As with all German Panzers that remained in production for an extended period, modifications were introduced to improve their performance. But with the exception of improved armor protection, there wasn't any significant difference in the performance of a **Pz.Kpfw.38(t)** from the first **Ausf.A** to the last **Ausf.G** - they all had the same drive train and weapons system. During the war the **Ausfuehrung** (model) designation was used only as an identifier to obtain the correct replacement parts. Without sig-

Panzerkampfwagen 38 (t)
Ausf.A to G and S
Fgst.Nr. Serie 1 - 1526

Weapons Data:

In Turret:	1 - 3.7 cm Kw.K.38(t)
	1 - 7.92 mm M.G.37(t)
Elevation:	-10, + 25 degrees
Traverse:	360 degrees
Gunsight:	T.Z.F.38(t) (2.6x, 25 degrees) graduated to 2000 meters
In Hull:	1 - 7.92 mm M.G.37(t)
Ammunition:	90 - 3.7 cm Pzgr
	2700 - 7.92 mm SmK

Crew: Commander/Gunner, Loader, Radio Operator, Driver

Communication: Fu 5 and Fu 2

Measurements:

Length, overall:	4.610 m
Width, overall:	2.135 m
Height, overall:	2.252 m
Firing Height:	1.710 m
Wheel Base:	1.770 m
Track Contact:	2.615 m
Combat Loaded:	9.725 (9.85) metric ton
Fuel Capacity:	220 liters

Automotive Capabilities:

Maximum Speed:	42 km/hr
Avg. Road Speed:	35 km/hr
Cross Country:	17 km/hr
Range on Road:	250 km
Cross Country:	100 km
Grade:	37 degrees
Trench Crossing:	1.90 m
Step:	80 cm
Fording Depth:	80 cm
Ground Clearance:	40 cm
Ground Pressure:	0.64 kg/cm^2
Power Ratio:	12.7 HP/ton
Steering Ratio:	1.48

Automotive Components:

Motor:	Praga Typ TNHPS/II, 6 cyl., water-cooled, 7.75 liter gasoline, 125 HP @ 2200 rpm
Transmission:	Praga-Wilson Typ CV
Reverse	6.1 km/hr
1. Gear	4.1 km/hr
2. Gear	10.3 km/hr
3. Gear	16.5 km/hr
4. Gear	26.2 km/hr
5. Gear	42.0 km/hr
Steering:	Clutch-brake
Drive:	Front sprocket
Roadwheels:	4 per side
Tires:	Rubber 775 mm dia.
Suspension:	Leaf springs
Track:	Dry pin 293 mm wide with 104 mm pitch
Links per side:	93

nificant automotive modifications, most replacement parts remained virtually unchanged during the entire production run; therefore the **Ausf.** identification of a **Pz.Kpfw.38(t)** merely reflects another contract extension and does not really identify a new, improved model.

However, there are several unique features that can be used to identify different **Ausf.** from each other, including:

Ausfuehrung A (Fgst.Nr. 1 to 150)

o The **Turmrundblickfernrohr 38(t)** (commander's observation periscope) head extended above the cylindrical armor guard and wasn't protected by an armor cap.

o A tubular **"battle" aerial** mounted on the left side was present on **Pz.Kpfw.38t Ausf.A** used in Poland by **Pz.Abt.67** but was discontinued before the end of the production run.

o A short-handled **shovel** and a **pickaxe** were stowed across the engine access hatch on the rear

o The **Laufwerk** (suspension) consisted of four roadwheels (with narrow rubber tires on thick steel bands) mounted in pairs with 14 leaf springs in each bundle without a slit in the side of the retainer.

o Initially, the front **reflectors** were mounted on the leading edge of each track guard and then relocated to behind the front support bracket.

o Starting in July 1939 and continuing to the end of the **Pz.Kpfw.38t Ausf.A** production run in November 1939, a **Nebelkerzenabwurfvorrichtung** (smoke grenade rack) without a protective armor guard was mounted at the left rear above the exhaust muffler.

o During the **Ausf.A** production run, a **stop** was bolted onto the turret front to prevent the ball-mounted machinegun from being traversed to the left toward the **3.7 cm Kw.K.38(t)**.

18-15

Panzerkampfwagen 38 (t) Ausf. A

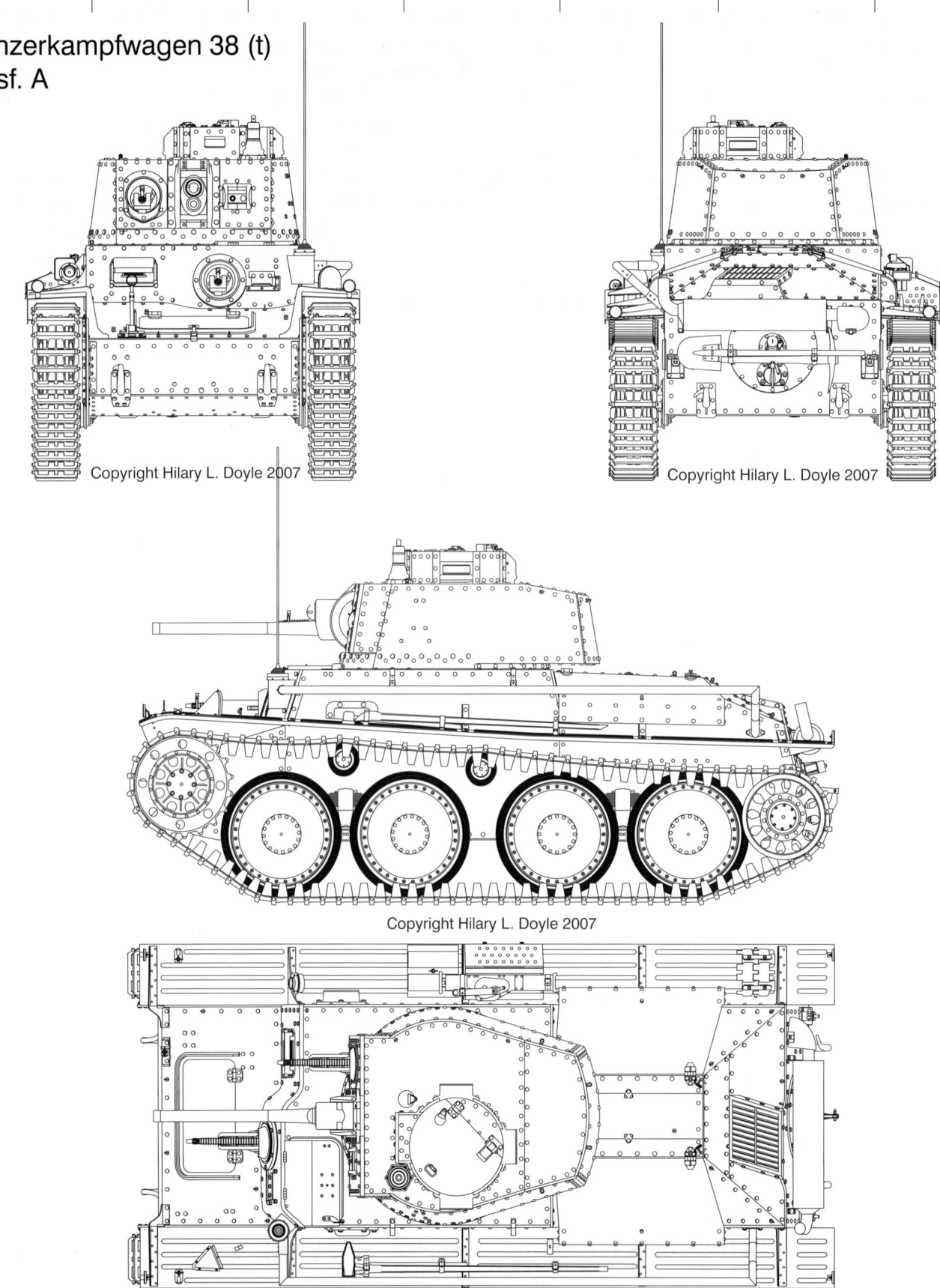

Features present at the start of the production run, included: the commander's observation periscope head extended above the cylindrical armor guard, both M.G.37(t) without flash hiders, a tubular "battle" aerial on the left side, a shovel and a pickaxe stowed on the hull rear, and narrow rubber tires with thick steel bands on the roadwheels.

o Toward the end of the production run, a **handle** was bolted onto the top left rear corner of the superstructure roof to assist the crew in mounting.

o A **Tarnscheinwerfer** (blackout light, commonly referred to as a "Notek") was mounted on the left track guard and an **Abstands-Ruecklicht** (convoy tail light) was mounted on the left rear (and relocated to the right rear) before the end of the production run.

Panzerbefehlswagen (Command Tanks)

Pz.Kpfw.38(t) could be outfitted with two racks for **Empfaenger e** (receiver sets), one rack for a **10 Watt Sender**, and a mount for three **Umformer** (rectifiers) so that they could be selectively used by the **Kompanie-Chef** (company commander with **Fu 5** and **Fu 2**), **Zugfuehrer** (platoon leader with **Fu 5**), or as a normal **Zugwagen** (platoon tank with **Fu 2**).

Pz.Kpfw.38(t) were converted to **Panzerbefehlswagen (Sd.Kfz. 266, 267, and 268)** at the Nachrichten Heereszeugamt, Berlin-Schoneberg (signals ordnance depot). Two **Sd.Kfz.266** for the **Abteilung Stab** were outfitted with **Fu 5** and **Fu 2** radio sets with a rod antenna. Two **Sd.Kfz.267** for the **Regiment Stab** and five **Sd.Kfz.267** for the **Panzerfunkkompanie a** were outfitted with **Fu 8** and **Fu 5** radio sets and a frame antenna on the rear deck. Starting with the **Ausf.B**, a dummy gun and mount replaced the **3.7 cm Kw.K.38(t)**, a vision port replaced the gun sight, and the turret was fixed in place on the **Pz.Bef.Wg.38(t)**

Panzerkampfwagen 38 (t) Ausf. A

Copyright Hilary L. Doyle 2007

Features present at the start of the production run, included: the commander's observation periscope head extended above the cylindrical armor guard, both M.G.37(t) without flash hiders, a tubular "battle" aerial on the left side, a shovel and a pickaxe stowed on the hull rear, and narrow rubber tires with thick steel bands on the roadwheels.

This Page and Upper Right: Panzerkampfwagen 38 (t) Ausf.A (Fgst.Nr. 3) as it was originally assembled by B.M.M. with the shovel and pick stowed on the rear. On all Pz.Kpfw.38(t) Ausf.A the head of the Turmrundblickfernrohr (periscope) extended above the armor guard. (BAMA)

Below: This Panzerkampfwagen 38 (t) (Fgst.Nr. 2) was tested by Wa Pruef 6 at Kummersdorf. The original tracks were worn out and have been replaced by the lighter cast Kettenglieder (track links) with a recess in the outer face of each guide tooth. (HLD)

(Sd.Kfz.267). Two **Sd.Kfz.268** for the **Panzerfunkkompanie a** were outfitted with **Fu 7** (for communication with aircraft) and **Fu 5** radio sets and rod aerials for both sets.

On 15 February 1940, the Korps-Kdo.XV wrote to A.O.K.4 requesting better command tanks: *The disadvantages of the Pz.Kpfw.35(t) and 38(t) as Fuehrer-Pz.Kpfw. are the small field of view of the optical devices, more difficult orientation, and more difficult observation, which makes command more difficult.* They wanted **Pz.Kpfw.III** and **gr.Pz.Bef.Wg.** in their place, but their request was rejected. On 27 February 1940, **Panzer-Abteilung 67** was ordered to send eight **Pz.Kpfw.38(t)** to the Nachrichten Heereszeugamt, Berlin-Schoneberg for conversion to **Panzerbefehlswagen**.

Ausfuehrung B (Fgst.Nr. 151 to 260)

o An armor cap, which rotated with the periscope, was added to protect the head of the **Turmrundblickfernrohr 38(t)** (commander's observation periscope).

o A sheet-metal **rain guard** was added above the sight aperture to prevent water from splattering onto the face of the telescopic gun sight.

o A **handle** to assist the crew in mounting was bolted onto the top of the turret roof on the left side.

o The **cylindrical antenna base** no longer had a pipe extension for attaching a "battle" aerial.

o **Tools** were stowed on both track guards (in a standardized pattern maintained to the end of the **Pz.Kpfw.38(t)** production run) with the shovel, pickaxe, and tow cables stowed on the left side.

o A **wider rubber tire** with a thin steel band for the roadwheels was introduced in the **Ausf.B** production run.

Ausfuehrung C (Fgst.Nr. 261 to 370)

As proven by a photograph of **Fgst.Nr.272**, this 12th **Pz.Kpfw.38(t) Ausf.C** still had the same features as the end of the **Ausf.B** production run with a rain guard for the gun sight, traverse stop for the turret machinegun, 25-mm-thick frontal armor, no turret ring guard, and tracks with taller/narrower guide teeth.

o While initially the **Ausf.C** still had 25-mm-thick frontal armor, **40 mm thick** armor on the hull front was introduced during the production run. The flanges securing these thicker front plates to the hull sides were redesigned for the thicker armor and had additional caps to protect the bolts used to secure the final drives.

o Also a three-piece **Abdeckung** (turret ring guard) to prevent rifle and machinegun fire from getting through the gap to the **Turmkugellager** (ball bearing race) was introduced during the production run.

o A **bump stop** was bolted to the hull side behind the first roadwheel to prevent damage to the suspension from bottoming out. The **Pz.Kpfw.38(t)** suspension oscillated and bottomed out, even when driven on smooth roads.

Ausfuehrung D (Fgst.Nr.371 to 475)

o As observed on a photo of **Fgst.Nr.377**, the spacing in the upper run of bolts on the "bent" superstructure front plate was changed with the **Ausf.D**. The 25-mm-thick plate in front of the driver was still 118 mm farther to the rear than on the left side where the **M.G.Kugelblende** (machinegun ball mount) was mounted.

o As proven by photos of **Fgst.Nr.377**, **379**, and **430**, 40 mm thick armor on the hull front was a standard feature on all **Ausf.D**.

o A German flexible rubber **Antennenfuss** in place of the Czech cylindrical antenna mount was introduced during the production run. The 2nd **Pz.Kpfw.38t Ausf.D**, **Fgst.Nr.377**, still had a Czech antenna base.

o Lighter cast **Kettenglieder** (track links) with a recess in the outer face of each guide tooth were introduced during the **Ausf.D** production run. The inner face of these wider/shorter guide teeth was sloped inward to help prevent throwing track. **Fgst.Nr.377** still had the original track with taller/narrower guide teeth.

Ausfuehrung E (Fgst.Nr. 476 to 750)

o This was the first **Pz.Kpfw.38t Ausf.** with a **straight driver's front plate**, thicker castings for both the driver's and radio operator's front visors, and a rectangular crew hatch in the front left corner of the superstructure roof.

o With the ends of the straight driver's front plate being mounted farther forward than the previous "bent" plate, the **first joint in the hull side plates** was now slanted forward to match the bottom of the front plate. The bump stop was then relocated to a position on the hull side behind the joint.

o **Zusatzpanzerung** (additional armor) was added to the turret front, superstructure front and sides, and hull front. Armor plates on the turret sides were increased to 30-mm-thick and 25-mm-thick on the turret rear. There were fewer bolts and rivets in a revised pattern holding the front plates.

o All **Ausf.E** had a flexible rubber **Antennenfuss** mounted on the left front corner of the superstructure.

o **Protective caps** were added on the rear to protect the track tension adjusters and center guide for the crank starter.

o Starting in March 1941, (trial vehicle **Fgst.Nr.624**) the exhaust muffler was raised to make room for mounting a **Nebelkerzenabwurfvorrichtung** (smoke grenade rack) in an armor guard in a protected location on the hull rear. The engine exhaust muffler was still mounted in its original lower position on **Fgst.Nr.622** and **634**.

o The suspension was reinforced by adding a **15th leaf spring** to the front bundle.

o During the **Ausf.E** production run, the positions of the **Abstands-Ruecklicht** (convoy taillight) and brake light were reversed, and later both lights were relocated to the sides and mounted on extension bars.

Two different versions of the Panzerkampfwagen 38 (t) Ausf.A als Pz.Bef.Wg. (modified as a command tank). As a Sd.Kfz.268 (above) with a rod antenna to communicate with aircraft and as a Sd.Kfz.267 (below) with a frame antenna for long range ground communication. (BA 128/419/04)

Panzerkampfwagen 38 (t) Ausf. B

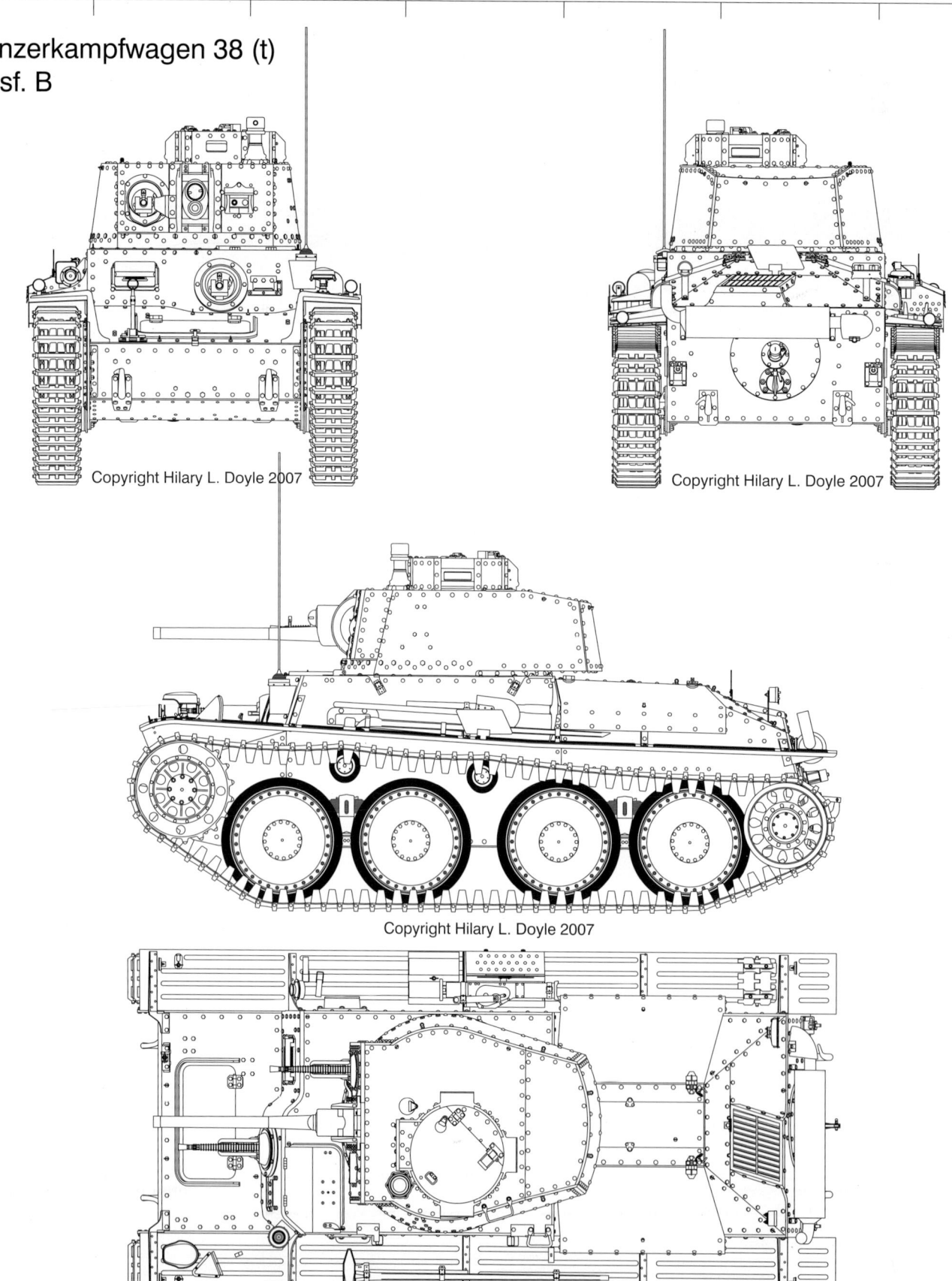

Features present on this Pz.Kpfw.38(t) Ausf.B include: an armor cap to protect the head of the commander's observation periscope, flash hiders on M.G.37(t), stop preventing turret machinegun from traversing left, rain guard for main gun sight, rhomboid shields for tactical numbers, handles on the turret roof and superstructure roof, standardized tool stowage, Notek blackout light, convoy tail light, and tail/brake light, wider rubber tires with thin steel bands on the roadwheels.

Above: A row of Panzerkampfwagen 38 (t) Ausf.B with Panzer-Regiment 25 of the 7.Panzer-Division have a rotating armor cap protecting the head of the Turmrundblickfernrohr (periscope) and flash-hiders on the M.G.37(t) machineguns. (NA)

Below: The tool and equipment stowage was rearranged on almost every Pz.Kpfw.38(t) after issue. (MJ)

Panzerkampfwagen 38 (t) Ausf. C

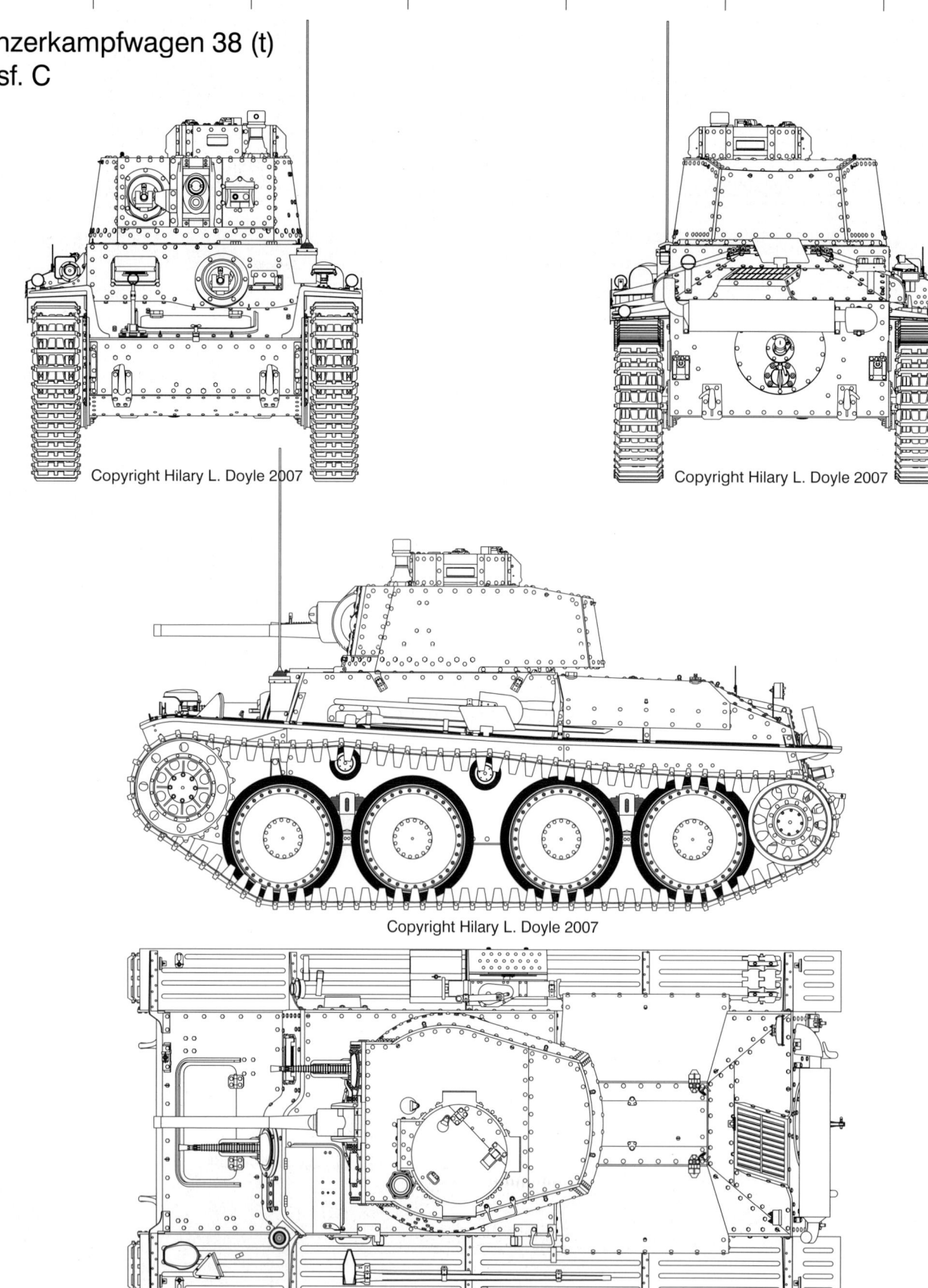

Copyright Hilary L. Doyle 2007

Features present on this Pz.Kpfw.38(t) Ausf.C include: 40 mm thick armor on the hull front with revised flanges at the hull sides and a bump stop mounted behind the first roadwheel.

18-24

Above: This brand new Panzerkampfwagen 38 (t) Ausf.C (Fgst.Nr. 272), completed in May 1940 and rushed to the front as a replacement, still has a 25 mm thick hull front and no turret ring guard. (KHM)
Below: A Panzerkampfwagen 38 (t) Ausf.C with a turret ring guard and 40 mm thick armor on the hull front. (KHM)

18-25

Panzerkampfwagen 38 (t) Ausf. D

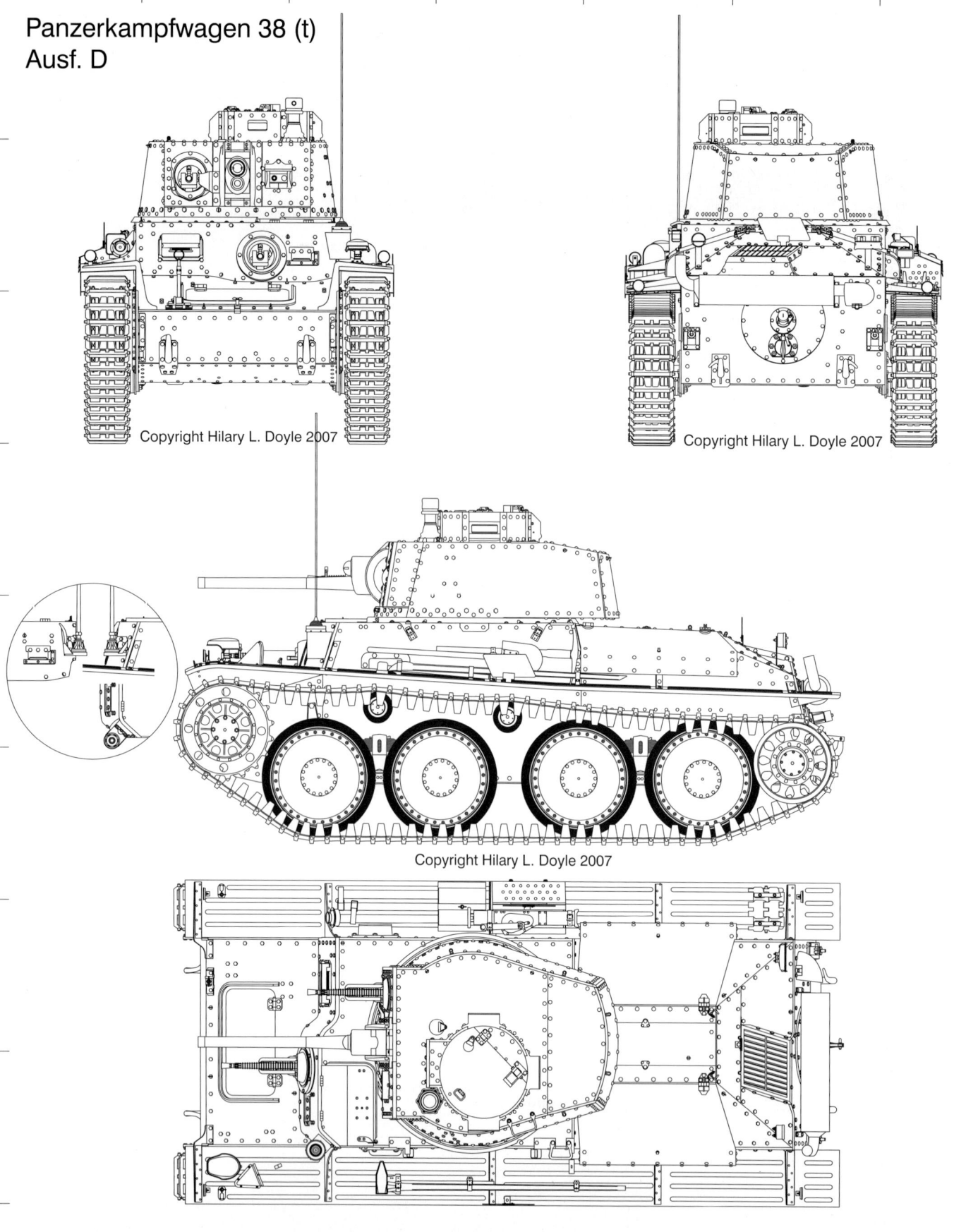

Features present on this Pz.Kpfw.38(t) Ausf.D include: revised spacing on the upper run of bolts on the "bent" superstructure front plate, 40 mm thick armor on the hull front with revised flanges at the hull sides, a three-piece turret ring guard, and a flexible rubber Antennenfuss (in circle) replacing the Czech aerial base.

Above: This brand new Panzerkampfwagen 38 (t) Ausf.D (Fgst.Nr. 377), completed in September 1940 with a turret ring guard, 25 mm thick driver's front plate, and 40 mm thick hull front, still has a Czech antenna base mount. (MJ)
Below: After being issued to Panzer-Regiment 21 in the 20.Panzer-Division the tool and equipment stowage have been rearranged and a rack for carrying "Jerry" cans added onto the rear deck. (KHM)

This Page and Upper Right: Panzerkampfwagen 38 (t) Ausf.D (Fgst.Nr. 379 (left) completed in September 1940 and Fgst.Nr. 430 (right) completed in October 1940) have already been modified by replacing the Czech antenna base with a German flexible rubber antenna base. Both fatter rolled hard copper 1.4 m rod antenna and thinner "whip" antenna with large base sockets were mounted. Tool and equipment stowage had been rearranged and ten spare track links secured on the glacis plate already by the time the photos were taken of these newly issued Pz.Kpfw.38 (t) in January 1941. (MJ)

Below: A new set of tracks with the lighter cast Kettenglieder (track links) with a recess in the outer face of each guide tooth. (MJ)

Panzerkampfwagen 38 (t) Ausf. E

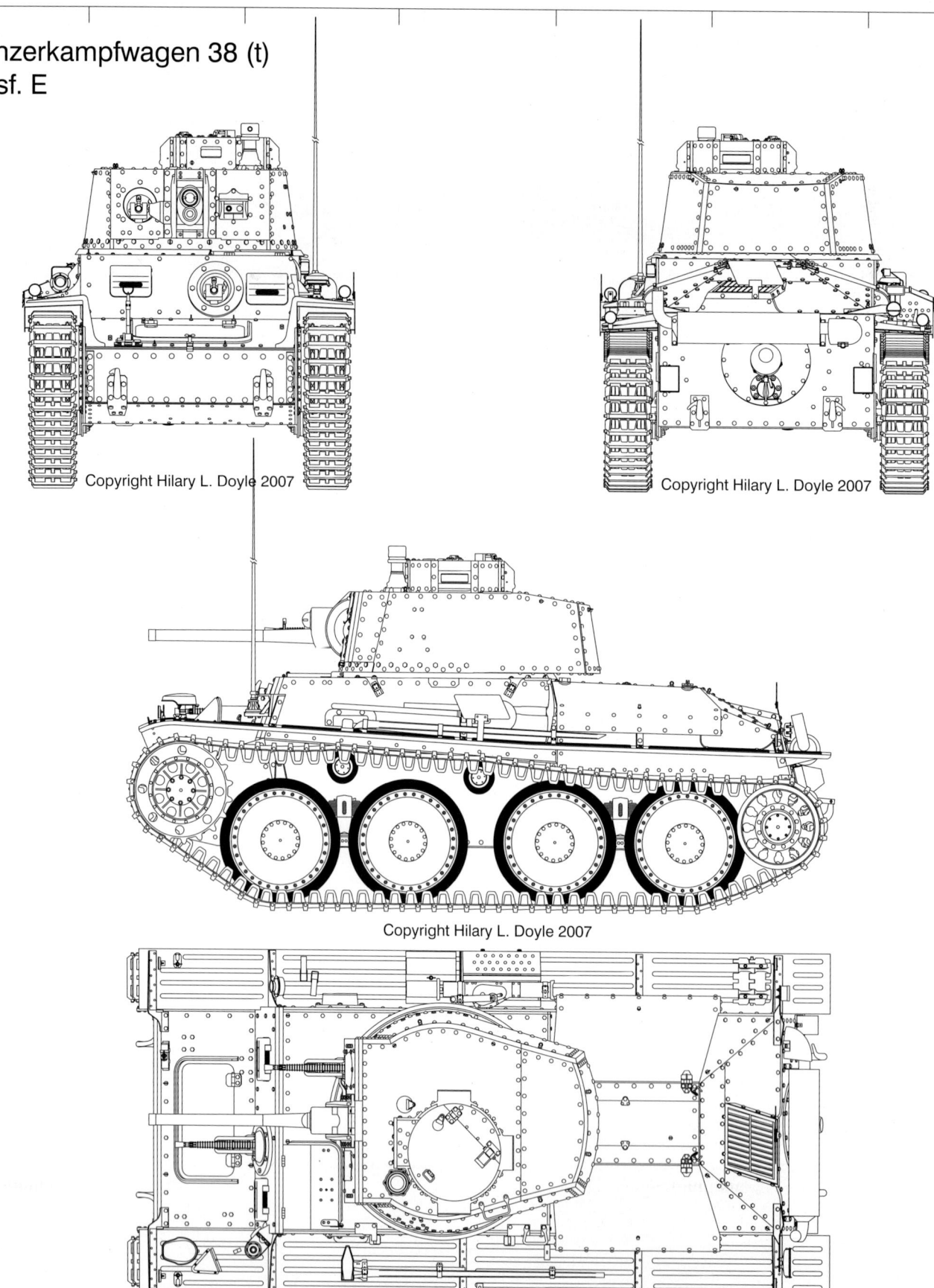

Features present on this Pz.Kpfw.38(t) Ausf.E include: 25 mm plus 25 mm Zusatzpanzer on the turret front, straight, superstructure front, and hull front, 15 mm plates bolted to the superstructure sides, thicker armor on the turret sides, rear, and roof, reduction in number of bolts/rivets holding armor, increased diameter turret ring guard, larger cast visors for the driver and radio operator, and protective caps on the track adjustor and crank starter guide.

Right:
Panzerkampfwagen 38 (t) Ausf.E (Fgst.Nr. 536) completed in January 1941 had increased armor protection with two 25 mm thick face-hardened plates bolted/rivetted together for the turret front, flat driver's front and hull front. An additional 15 mm thick plate was bolted onto the superstructure side but not to the hull side below the track guards. The armor thickness was increased to 30 mm on the turret sides and 25 mm on the turret rear. (MJ)

Below:
This Pz.Kpfw.38(t) Ausf.E with a raised muffler has been outfitted with a tow hitch and quick release mechanism for a Betriebstoff-Anhaenger (Fassungsvermoegen 200 l) (fuel trailer with 200 liter drum). (KHM)

Panzerkampfwagen 38 (t) Ausf. F

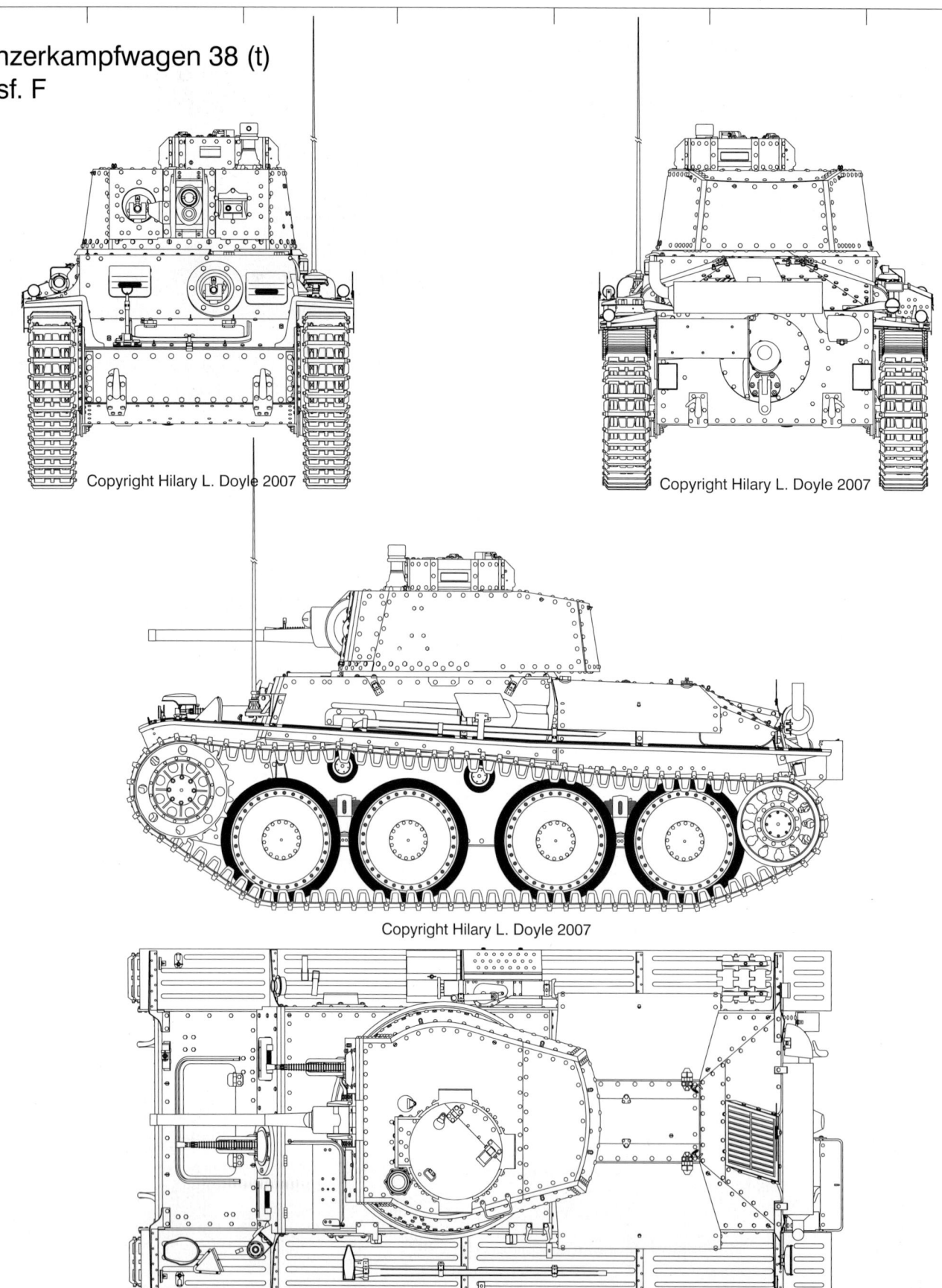

Copyright Hilary L. Doyle 2007

Features present on this Pz.Kpfw.38(t) Ausf.F include: engine exhaust muffler raised to make room for a smoke grenade rack with an armor guard to be located in a protected location on the hull rear, reduced number of rivets on the engine hatches, three spare track links on the right rear track guard, and a tow coupling for a fuel trailer.

Right:
A brand new Panzerkampfwagen 38 (t) Ausf.F (Fgst.Nr. 794) completed in June 1941 still has the mirror and Notek Tarnscheinwerfer mounted on the left track guard.
(MJ)

Below:
Still at the BMM assembly plant, this Pz.Kpfw.38(t) has been modified by stowing 10 spare track links on the glacis plus 8 track links held by two bars across the hull front.
(CKD)

18-33

This and Opposite Page: A Panzerkampfwagen 38 (t) Ausf.F still at the BMM assembly plant has been altered by experimentally stowing seven spare track links on the hull front. This "standard" modification was later introduced into the Pz.Kpfw.38(t) production series starting with Ausf. G Fgst.Nr.1220. Most of the bolts along the sides of the engine compartment hatch have been dropped. It still has the tow hitch for the fuel trailer but doesn't have the quick release mechanism. Five smoke grenades which could be released singly were mounted on the hull rear and protected from being set off by small arms fire by an armor guard. (CKD)

18-35

This Page:
With the exception of three instead of two spare track links on the right rear track guard and ten spare track links on the glacis plate, this Pz.Kpfw.38(t) Ausf.F at the BMM assembly plant still has the standard tool and equipment stowage arrangement introduced with the Pz.Kpfw.38(t) Ausf.B with the jack, jack block and tool box on the right track guard, and the shovel, pick, tow cable, and wrecking bar on the left. (CKD)

Ausfuehrung F (Fgst.Nr. 751 to 1000)
o All Ausf.F still had **Zusatzpanzerung** (additional armor) on the turret front, superstructure front and side, and hull front with the same bolt and rivet pattern as an **Ausf.E**.
o Three (instead of two) **spare track links** were stowed toward the rear on the right track guard.
o The number of bolts along the side of the **Motorabdeckklappe** (engine hatches) was reduced during the **Ausf.F** production run.
o During the production run the **Tarnscheinwerfer** (Notek) was relocated to a base mounted on the left front corner of the glacis plate (after **Fgst.Nr.794**, prior to **Fgst.Nr.814**).

Ausfuehrung S (Fgst.Nr. 1001 to 1090)
Initially designed before the **Ausf.E,** but assembled at the same time as the **Ausf.F,** resulted in a mixture of features on the **Ausf.S**.
o With the ends of the straight driver's front plate mounted farther forward than the previous "bent" plate, the **first joint in the hull side plates** was now slanted forward to match the bottom of the front plate. The bump stop was then relocated to a position on the hull side behind the joint.
o 25 mm thick **Zusatzpanzerung** (additional armor) was added to improve armor protection but **only** on the turret, superstructure, and hull front - not on the turret or hull sides, which were still 15 mm thick.
o There were six bolts along the top edge of the driver's front plate above the original driver's visor, which had been designed for mounting onto a single 25-mm-thick plate. The smaller radio operator's visor was also retained, but without three bolt heads on its face and the splash guard rim.
o The **Ausf.S** had the same bolt and rivet pattern (12 in the top row) on the face of the hull front plate as the **Ausf.A** to **D** but with a flange secured with five bolts at both ends next to the hull side.
o The **rivet and bolt pattern** on the side of the "box" extension on the turret front was changed to **five rivets in two rows** - instead of six in two rows on the **Ausf.A** to **D** and four in two rows on the **Ausf.E** to **G**.
o Since the side of the turret was still 15 mm thick, the **turret ring guard** was still the same size that was mounted on the **Ausf.C** and **D**.
o **Protective caps** were added on the rear to protect the track tension adjusters and center guide for the crank starter.
o The exhaust muffler was raised to make room for mounting a **Nebelkerzenabwurfvorrichtung** (smoke grenade rack) in an armor guard, located in a protected location on the hull rear.

Ausfuehrung G (Fgst.Nr. 1100 to 1359, 1480 to 1526)
o Armor protection on the **Ausf.G** was upgraded to **50-mm-thick homogeneous armor plates** on the turret, superstructure, and hull front.
o The number of rivets and bolts was reduced on the turret and superstructure front.
o A rack for **Ersatzkettenglieder** (spare track links) mounted on the hull and straps to hold five spare track links on the right and left side of the glacis plate were introduced starting with **Fgst.Nr.1220** in November 1941.
o The **mirror** mounted on the left track guard was dropped during the **Ausf.G** production run.
o Starting in February 1942, a **Delbag-Luftfilter** (air filter) replaced the **Doppelluftfilter** (double air filter). With the installation of the improved air filter, engine combustion air was drawn from the **Kampfraum** (fighting compartment).
o An adjustable **Schieber** (sliding baffle plate) was mounted

Right: A Panzerkampfwagen 38 (t) Ausf.S with a raised muffler and caps protecting the track adjusters and crank starter. (BA 213/276/14)

18-37

Panzerkampfwagen 38 (t) Ausf. S

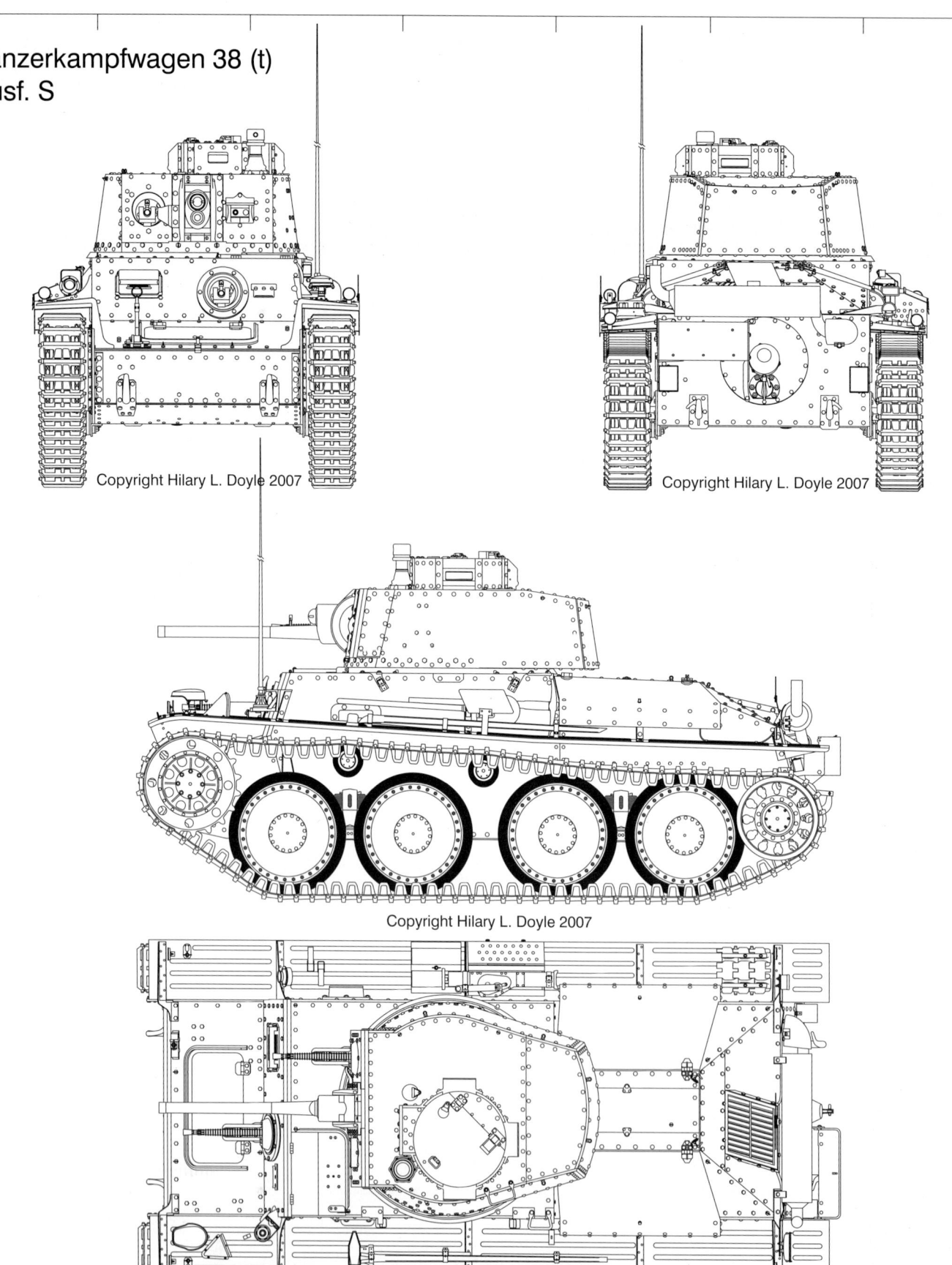

Copyright Hilary L. Doyle 2007

Features present on this Pz.Kpfw.38(t) Ausf.S include: forward joint in hull side cut at a slant to match a straight driver's front plate, 25 mm Zusatzpanzer bolted onto the turret front, straight driver's front plate, and hull front (but all side armor still 15 mm thick), original driver's/radio operator's visors, reduced number of bolts/rivets, same turret ring guard as Ausf.C/D, raised engine exhaust muffler, smoke grenade rack, and protective caps on track adjustors and crank starter guide.

18-38

This Page: A Panzerkampfwagen 38 (t) Ausf.S, identified by two rows of five bolts on the front extension of the turret side, did not have increased armor protection on the turret sides and rear or on the superstructure/hull sides. (BA 213/267/12 & 13)

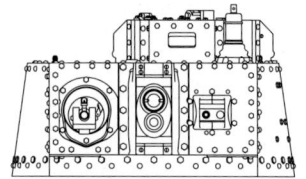 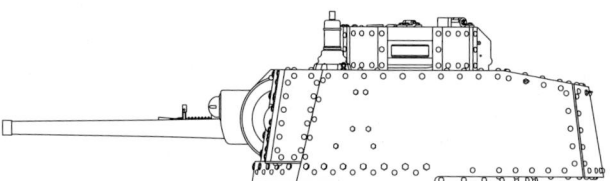 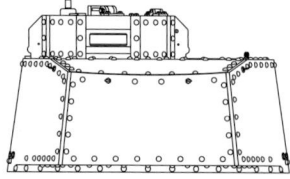

Pz.Kpfw.38(t) Ausf.A Turm with 25 mm front, 15 mm side, and 15 mm rear armor
Identifiable by the rivet pattern on the front and forward side and no armor cap on top of the periscope armor guard

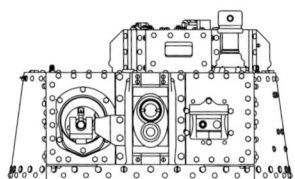 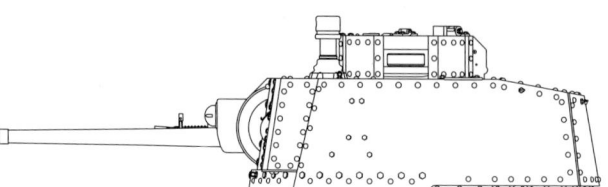 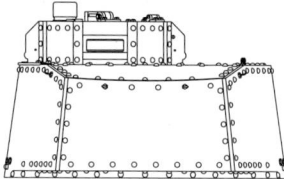

Pz.Kpfw.38(t) Ausf.B, C and D Turm with 25 mm front, 15 mm side, and 15 mm rear armor
Identifiable by the rivet patten on the front and foward side and armor cap for the head of the observation periscope

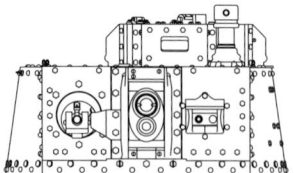 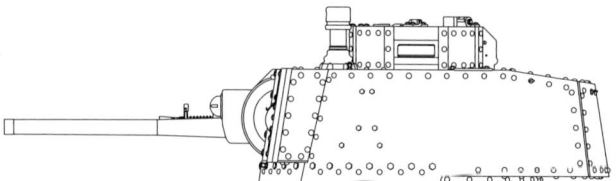 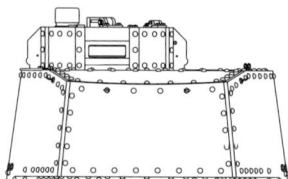

Pz.Kpfw.38(t) Ausf.S Turm with 25 mm + 25 mm front, 15 mm side, and 15 mm rear armor
Identifiable by the rivet pattern on the turret front and two rows of five rivets on the turret side forward extension

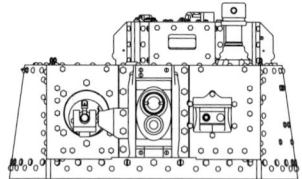 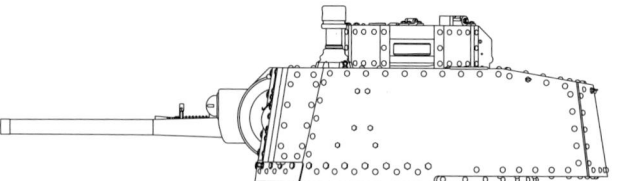 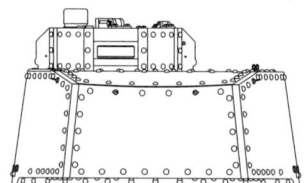

Pz.Kpfw.38(t) Ausf.E/F Turm with 25 mm + 25 mm front, 30 mm side, and 25 mm rear armor, and handle rivetted on roof
Identifiable by the rivet pattern on the turret front and two rows of four rivets on the turret side forward extension

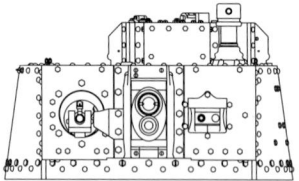 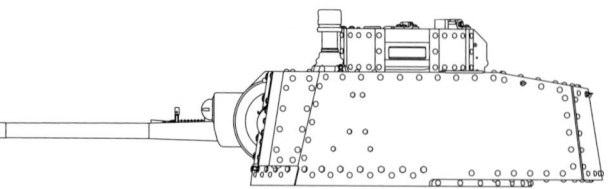 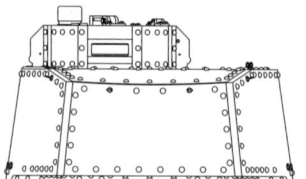

Pz.Kpfw.38(t) Ausf.G Turm with 50 mm front (except 25 mm lower curve), 30 mm side, and 25 mm rear armor
Identifiable by the rivet pattern on the turret front and two rows of four rivets on the turret side forward extension

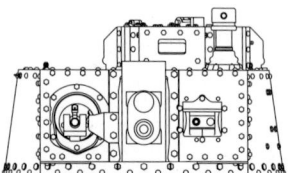 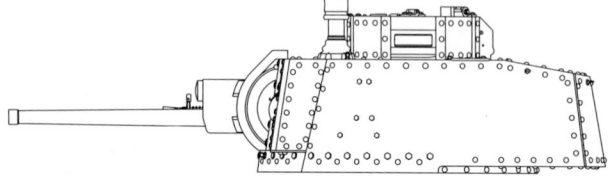 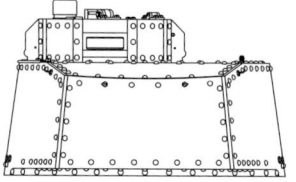

Pz.Kpfw.38(t) Ausf.B Turm converted to a Panzerbefehlswagen with a dummy gun

Copyright Hilary L. Doyle 2007

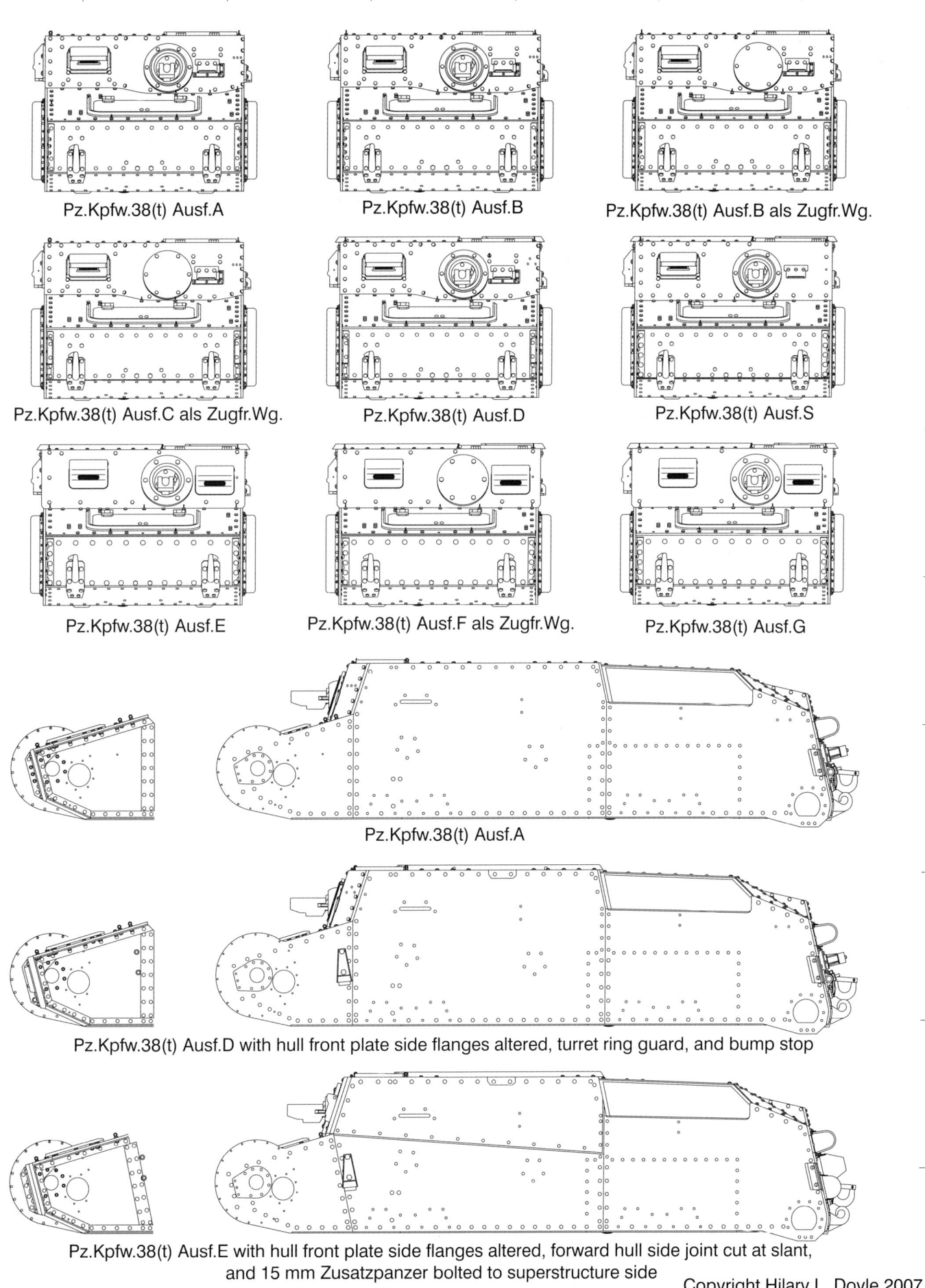

Panzerkampfwagen 38 (t) Ausf. G

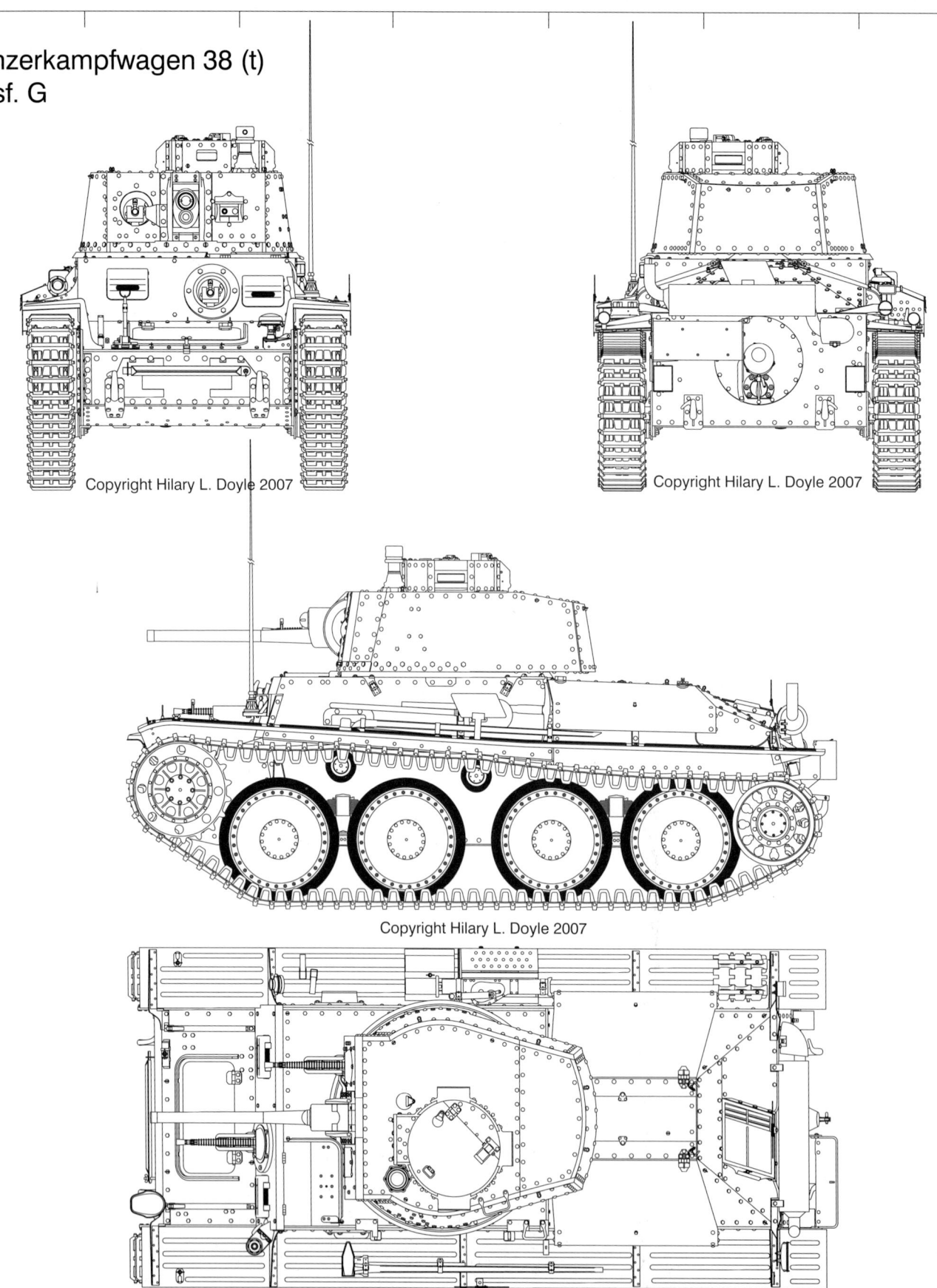

Copyright Hilary L. Doyle 2007

Features present on this Pz.Kpfw.38(t) Ausf.G include: 50 mm thick homogenous front plates on the turret, superstructure, and hull, reduced number of bolts/rivets holding armor, a slide added to the engine air exhaust opening to regulate cooling air flow, a rack and straps for spare track links on the glacis and hull front, relocated Notek black out light, reinforcing caps welded over the slot in spring bundle retainers, and a maintenance access hatch in the belly below the engine.

18-42

over the cooling air exhaust opening and used to regulate engine temperature, replacing the internally adjustable louvers in the firewall of previous **Ausf.**

As described in a report dated 30 January 1942, the following changes were needed to convert a **Pz.Kpfw.38(t)** into a **Tropenfaehige Pz.Kpfw.** (tank suitable for operations in hot climates):

1. To improve cooling air flow, replace the louvers in the air intake duct with improved ones.
2. To regulate the cooling air flow, install an adjustable **Schieber** (sliding baffle) on the air exhaust opening.
3. To lower the temperature in the driver's and radio operator's positions, install two suction air ducts.
4. For protection in a sandstorm, outfit the **Pz.Kpfw.** with a tarp to cover the engine compartment and add the necessary fasteners. This will be replaced with a canvas cover for the entire vehicle, which is currently being developed for the **Pz.Kpfw.38(t)**.
5. To improve engine oil filtration, install an oil filter.
6. To ventilate the electric generator, protect the sieve filter from dust.
7. Outfit the vehicle with two **Kanistern** (Jerry cans) for distilled and drinking water.

In May 1942, one **Pz.Kpfw.38(t) Ausf.G, Fgst. Nr.1481**, was outfitted as a **Tropen Pz.Kpfw.** for trials.

Panzerkampfwagen 38 (t) Ausf. G

Copyright Hilary L. Doyle 2007

Features present on this Pz.Kpfw.38(t) Ausf.G include: 50 mm thick homogenous front plates on the turret, superstructure, and hull, reduced number of bolts/rivets holding armor, a slide added to the engine air exhaust opening to regulate cooling air flow, a rack and straps for spare track links on the glacis and hull front, relocated Notek black out light, reinforcing caps welded over the slot in spring bundle retainers, and a maintenance access hatch in the belly below the engine.

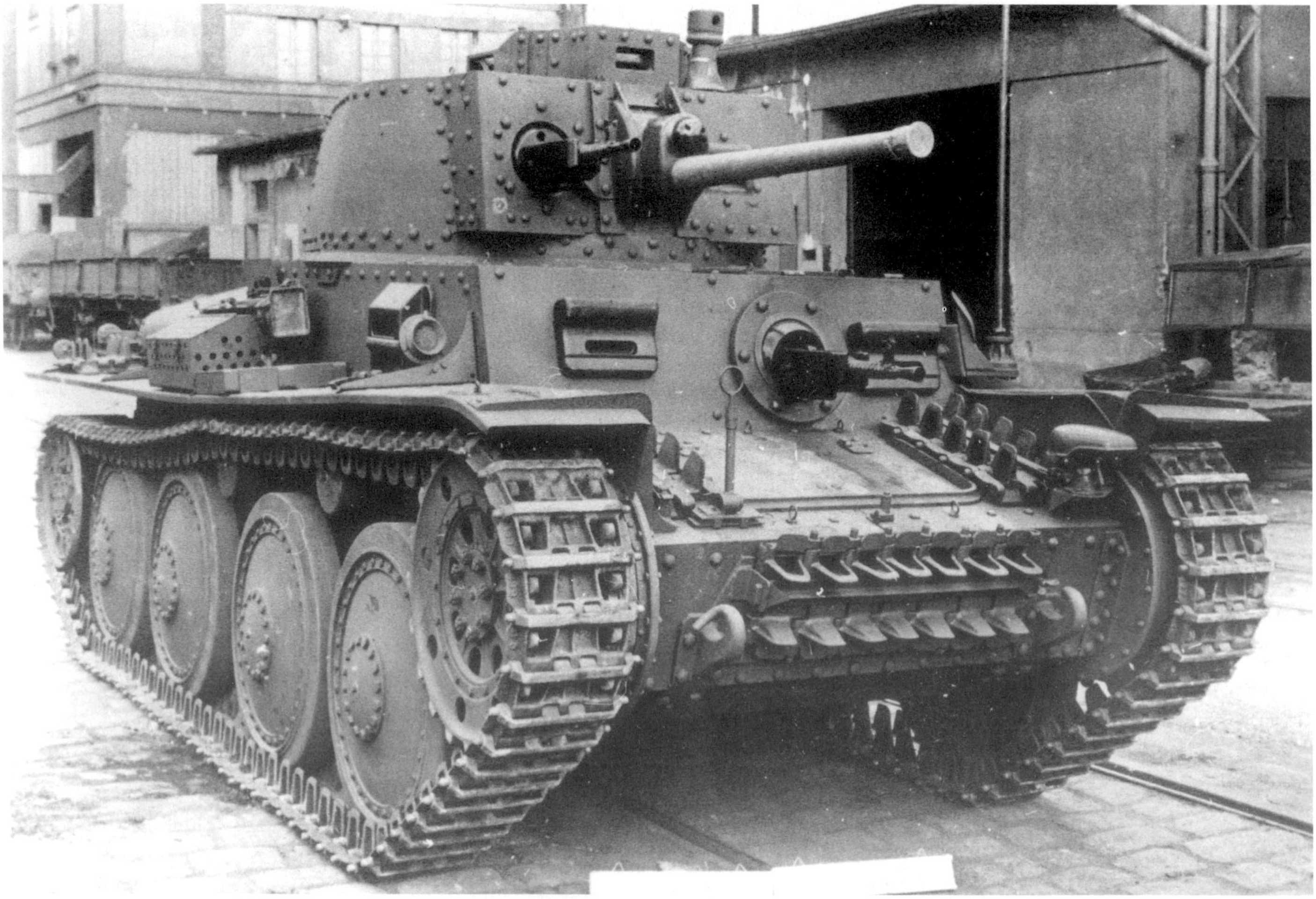

18-44

This Page and Opposite: A Panzerkampfwagen 38 (t) Ausf.G, photographed at the BMM assembly plant on 2 June 1942, had reinforced brackets for the leaf spring bundles, the mirror dropped off the right track guard, and the Notek Tarnscheinwerfer relocated to the left side of the hull front. (CKD)

18-45

Backfitted Modifications

After being issued to units, each Panzer-Regiment "improved" the external stowage on the **Pz.Kpfw.38(t)** by adding stowage boxes and racks and rearranging tool stowage on the track guards and securing spare track links on the front.

The Notek **Tarnscheinwerfer** and associated **Abstands-Ruecklicht** were backfitted to the **Pz.Kpfw.38(t) Ausf.A**.

In a modification order dated 20 October 1940, rubber fuel lines in the **Pz.Kpfw.38(t) Ausf.A** were replaced with fuel lines resistant to being "eaten" by gasoline.

A flexible rubber **Antennenfuss** mounted on a sheet metal shelf replaced the Czech cylindrical antenna mount housing a spring to pull a tipped antenna back upright. **Pz.Kpfw.38t Ausf.D Fgst.Nr.377**, assembled in September 1940, still had a cylindrical Czech antenna base. A thin "whip" antenna with a wider socket at the base was used infrequently as a substitute for the thicker rod antenna (made out of rolled hard copper) mounted on the flexible rubber **Antennenfuss**.

When their tracks wore out, **Pz.Kpfw.38(t) Ausf. A**, **B**, **C**, and **D** were refitted with the newer track links with wider/shorter guide teeth and **Pz.Kpfw.38(t) Ausf.A** with roadwheels with wider rubber tires.

On 30 May 1941, OKH, Chef H. Ruest u. B.d.E. ordered that every **Pz.Kpfw.35(t)** and **38(t)** and **Pz.Bef. Wg.35(t)** and **38(t)** be outfitted immediately with a **Betriebstoff-Anhaenger (Fassungsvermoegen 200 l)** (fuel trailer with 200 liter drum) and a **Ueberpumpvorrichtung** (transfer pump device) to increase their driving range.

On 10 January 1942, AHA/Ag K/In 6 authorized a modification to add a holder for **Ersatzkettenglieder** (replacement track links) on **Pz.Kpfw.38(t) Fgst.Nr.1 to 1219**. The needed parts were to be made by the unit in accordance with requisitioned modification drawings.

Pz.Kpfw.38(t), sent back for major overhaul, were upgraded with the "latest" modifications, including raised engine exhaust mufflers and protective caps on the rear to protect the track tension adjusters and center guide for the crank starter.

Above: The quick release tow hitch for the Betriebstoff-Anhaenger (Fassungsvermoegen 200 l) (fuel trailer with 200 liter drum) and Ueberpumpvorrichtung (transfer pump device) modification added to increase the driving range of Pz.Kpfw.38(t). (CKD)

This Page: An experimental attempt to mount the Nebelkerzenabwurfvorrichtung (smoke grenade rack) with protective armor cover onto a Panzerkampfwagen 38 (t) Ausf.A (Fgst.Nr. 69) with the original muffler configuration. (CKD)

1939 Campaign in Poland

At the start of the campaign in Poland on 1 September 1939, the **3. leichte Division** was reported to have 55 **Pz.Kpfw.III(t)** and 2 **Pz.Kpfw. III(t) als Bef.Wg.** in **Pz.Abt.67**. The total number available in the entire army inventory was reported as 57 with the **Feldheer** on 31Aug39, 36 stockpiled in the **Heereszeugamt** on 30Sep39, and 5 with the **Ersatzheer** (reserves) including **Schule,** for a total of 98 **L.T.M.38**.

After the campaign the **3.leichte Division** reported that 7 out of the 57 **L.T.M.38** had been lost as unrepairable total writeoffs during the period from 1 to 25 September 1939.

This and Opposite Page: Pz.Kpfw.38(t) Ausf.A with Pz.Abt. (verlastet) 67 in the 3.leichte Division before (left), during (right), and after (below) the campaign in Polen in 1939. "Verlastet" meant transported on trucks and trailers as an attempt to increase the speed in which the unit could advance or be shifted to a different sector. The experiment wasn't repeated after this campaign. The Panzer-Kompanien in Pz.Abt.(verl.) 67 were numbered 2., 3., and 4.
(3 x MJ, 1 HLD)

1940 Campaign in the West

As reported on 5 February 1940, the **8.Panzer Division** lacked only 4 **Pz.Kpfw.38(t)** in the **3.Kp./Pz.Rgt.10** and 6 in their **Staffel** (reserve) from being completely outfitted to their authorized strength of 129 **Pz.Kpfw.38(t)** and 15 **Pz.Bef.Wg.** (including 7 with the **2.Funk-Kompanie/Panzer-Nachrichten-Abteilung 84**). At this time the **3.Kp./Pz.Abt.67** still had 17 **Pz.Kpfw.38(t),** but subsequently this company was converted into a **mittlere Kompanie** and issued **Pz.Kpfw.IV** before the offensive started.

The next units to be issued **Pz.Kpfw.38(t)** were **Panzer-Regiment 25** and **Panzer-Abteilung 66** in the **7.Panzer-Division**. As ordered on 21 February 1940, they were issued a total of 30 **Pz.Kpfw.38(t)**, five for each of their six **leichte Panzer-Kompanien,** and ordered to turn in the previously issued 10 **Pz.Kpfw.III**. As authorized by orders dated 6 March 1940, each **leichte Panzer-Kompanie** was to be organized in accordance with **K.St.N.** and **K.A.N.1171(Sd)** dated 1Sep39. However, until they received additional **Pz.Kpfw.38(t)** they were to be organized with the 5 **Pz.Kpfw.38(t)** consolidated into one **Zug** (platoon) within each **leichte Panzer-Kompanie**.

As additional **Pz.Kpfw.38(t)** became available, they were issued to units in the **7.Panzer-Division** as follows:

o 10 **Pz.Kpfw.38(t)** including 2 **Zugfuehrerwagen** issued to **Pz.Abt.66** from HZA Magdeburg on 7 March

o 31 **Pz.Kpfw.38(t)** arrived on 6 April

o 2 **Pz.Bef.Wg.Sd.Kfz.267 auf Fgst.38(t)** issued to **Pz.Rgt.25** on 14 April

o 5 **Pz.Kpfw.38(t)** including 1 **Zugfuehrerwagen** issued to **Pz.Abt.66** from HZA Magdeburg on 23 April

o 10 **Pz.Kpfw.38(t)** including 2 **Zugfuehrerwagen** issued to **Pz.Rgt.25** on 30 April

o 7 **Pz.Kpfw.38(t) als Bef.Wg.** issued to **Pz.N.Abt.83** on 2 May

o 5 **Pz.Kpfw.38(t)** including 1 **Zugfuehrerwagen** issued to **Pz.Rgt.25** from HZA Magdeburg on 6 May

Panzer-Regiment 25 had started out on 10 May with 84 operational **Pz.Kpfw.38(t)** including **Pz.Bef.Wg.** and still had 49 operational on 26 May 1940, with 9 written off as total losses. Two additional **Pz.Kpfw.38(t)** had been pulled forward from their previous garrison in Meckenheim. 50 **Pz.Kpfw.38(t)** were still operational on 29 May. On 30 May, the **7.Panzer-Division** reported that they had knocked out 18 heavy and 295 light tanks, of which 6 heavy and 171 light tanks were credited to **Panzer-Regiment 25** for the total loss of 20 **Pz.Kpfw.38(t)**, 4 **Pz.Kpfw.IV**, 9 **Pz.Kpfw.II**, and 10 **Pz.Kpfw.I**.

The spirit in which the Panzers were handled is revealed in the following excerpts from a citation signed by Generalmajor Rommel recommending Hauptmann Adalbert Schulz, **Chef 1.Kompanie/Panzer-Regiment 25**, be awarded the **Ritterkreuz** (knight's cross):

*After the **1.Kp./Pz.Rgt.25** had crossed to the west bank of the Maas 2-1/2 kilometers north of Dinant with four Pz.Kpfw.II and seven Pz.Kpfw.III(t) operational, on 14 May 1940 at 0800, Hptm. Schulz, Kp.-Chef received an order from the regimental commander, Oberst Rothenburg, to quickly go to the aid of the Schtz.Rgt.6 on the heights west of the Maas which was heavily pressured by enemy tanks. After reaching the heights by Grange, without pausing Hptm. Schulz attacked and threw back the strong enemy force in the woods west of Grange. The **Kompanie** destroyed two French tanks.*

*In spite of heavy French artillery fire, **Komp. Schulz** cleaned up the woods west of Grange to its south edge and drove the enemy back toward Haut le Wastia. While feeling forward toward Hontier, **Komp. Schulz** was fired at by 6 to 8 French anti-tank guns. Three Pz.Kpfw.III(t) and two Pz.Kpfw.II fell out. **Komp. Schulz** received orders from the Regiment to break off the action and move to the Regiment by the crossing point east of Grange.*

During the march there, Hptm. Schulz learned that to his rear about 14 French tanks were attacking from the direction of Sommiere. Even though he still had only three Pz.Kpfw.III(t) and two Pz.Kpfw.II, Hptm Schulz again turned around and hurried to help the threatened Schuetzen.

On the edge of the wood by Grange, Hptm. Schulz took on the fight with the numerically superior and heavier (one 32 ton, several 16 ton) enemy tanks. Another two enemy tanks were knocked out; the others pulled back toward Sommiere after about a 20-minute firefight. Afterward, Komp. Schultz returned to the Regiment.

*On 15 May, with the remainder of the **I.Abt./Pz.Rgt.25** reorganized into the **2.Kp./Pz.Abt.66**, Hptm. Schulz led the tank attack from Anthee through Philippeville to Cerfontaine. There **Komp. Schulz** received the order to drive back to Philippeville with*

Above:
A Pz.Kpfw.38(t) being loaded on board a ship purportedly one of 15 that were ordered to be sent to Norway with three independent Panzer-Zuege. (NA)

Right:
The last Pz.Kpfw.38(t) in this column is an Ausf.A (without an armor cap on the periscope). There is a white aerial recognition panel painted on the rear decks of these Pz.Kpfw.38(t) with Panzer-Regiment 25 in the 7.Panzer-Division in 1940. (MJ)

*the division commander in order to gain contact with the main body of the division. While the Panzer attack was conducted to the north past Philippeville, **Komp. Schulz** went around Philippeville to the south during the return trip from the division commander. Next **Komp. Schulz** took numerous French prisoners. Then the **Panzer-Kompanie** suddenly encountered an enemy column guarded by tanks and anti-tank guns about 1-1/2 kilometers southwest of the city of Philippeville. Hptm. Schulz immediately attacked. Three enemy tanks were knocked out, two anti-tank guns were rendered unusable, and part of the column surrendered.*

*Due to detachment for prisoner transport and losses from mechanical failures, **Komp. Schulz** was now only at platoon strength.*

*South of Phillipeville, with his platoon Hptm. Schulz suddenly ran into three French 18 ton tanks behind a curve. Opening fire at such short range was no longer possible. The French had already aimed their guns at the platoon. Courageously Hptm. Schulz drove swiftly between the French tanks and demanded that the French surrender. First the French tank commanders standing outside the tanks fired at Hptm. Schulz with pistols. Alone, by rashly charging after a short fight, Hptm. Schulz managed to get the French with their tanks to surrender. Furthermore he captured another 10 light French tanks with their crews that stood directly behind the first three with their engines running. Shortly afterward, **Komp. Schulz** managed to gain contact with **Schtz.Brig.7**.*

*On 23 May, **Komp. Schulz** received orders from the regiment to depart Barlin and within the regiment's sector take over guarding Bonvigny. **Kompanie** strength was two platoons each with four **Pz.38(t)**.*

*While moving through Hersin, Hptm. Schulz was informed by retreating **Schuetzen** and **Flak** elements of an ongoing enemy tank attack on the north side of Hersin with at least 10 tanks.*

*Hptm. Schulz immediately set his **Kompanie** against the enemy tanks that had already penetrated into the **Schuetzen** position. Schulz and his **Kompanie** attacked without pausing, destroyed three of the enemy tanks, and sent the rest into flight. After our own **Artillerie** went into positions on the north edge of Hersin and the situation was secure, Hptm. Schulz moved to fulfill the task assigned by the Regiment's orders.*

On 2 June 1940, OKH Gen.d.H. Gen.Qu ordered that 20 **Pz.Kpfw.38(t)** be issued to the **7.Panzer-Division** as replacements. At the end of the campaign on 29 June 1940, the **7.Panzer-Division** reported that **Panzer-Regiment 25** had started the campaign with 91 **Pz.Kpfw.38(t)**, 6 **Pz.Bef.Wg.(t) (Sd.Kfz.266)**, and 4 **Pz.Bef.Wg.(t) (Sd.Kfz.267)**, had received 21 **Pz.Kpfw.38(t)** as replacements, and had total losses of 25 **Pz.Kpfw.38(t)** and 2 **Pz.Bef.Wg.(t) (Sd.Kfz.267)**. There were still 59 **Pz.Kpfw.38(t)** operational on 1 July 1940.

Panzer-Regiment 10 with the **8.Panzer-Division** had started the offensive with 116 **Pz.Kpfw.38(t)** and 15 **Pz.Bef.Wg.(t)** on 10 May 1940. They reported still having 62 and 15 **Bef.** operational on 29 May, which had increased to 105 and 9 **Bef.** operational by the start of the second phase on 3 June 1940. On 4 June 1940, **Panzer-Regiment 10** was ordered to pick up 16 **Pz.Kpfw.38(t)** replacements that were available from Gruppe Thomale in Mons.

Near the end of the campaign on 20 June, there were only 16 newly produced or factory-repaired **Pz.Kpfw.38(t)** available as replacements in H.Za. Magdeburg. But these were not being issued and sent to the front, because timely delivery couldn't be made due to the shortage of rail transport, and further breakdowns would occur on long road marches. By 21 June 1940, a total of 8 **Pz.Kpfw.38(t) (Fgst.Nr.89/130, 91, 99, 113, 114/150, 126/76, 139, and 179/36/2)** had been loaded on rail cars to be returned to the homeland for major repair. On 10 July 1940, it was reported that the small stockpile of Panzers available in the Reich fell far short of being able to cover the losses during the campaign. 38 **Pz.Kpfw.38(t)** which were available on 10 July were issued by the end of July to refresh the Panzer-Divisions. With this supply, the Panzer-Divisions were restored to about 80% of their operational strength that they had had at the start of the campaign on 10 May 1940. On 22 October 1940, **Panzer-Regiment 10** was overstrength at 17 to 22 per **leichte Panzer-Kompanie** and a total of 127 **Pz.Kpfw.38(t)**.

Above:
A Pz.Kpfw.38(t) Ausf.A with Panzer-Regiment 10 of the 8.Panzer-Division during the campaign in the West in 1940.
(MJ)

Right:
A Pz.Bef.Wg.38(t) in the 2.Kompanie of Panzer-Nachrichten-Abteilung 83 in the 7.Panzer-Division after the campaign in the West in 1940.
(KHM)

1941 Campaign in the East

Three additional Panzer-Divisions were formed in the Fall of 1940 with Panzer-Regiments outfitted with **Pz.Kpfw.38(t)** - the **12.Panzer-Division** with **Panzer-Regiment 29**, the **19.Panzer-Division** with **Panzer-Regiment 27**, and the **20.Panzer-Division** with **Panzer-Regiment 21**. These, along with the experienced **7.** and **8.Panzer-Division**, were sent into action on 22 June 1941 with the strengths shown in the following table.

Operational Status on 22 June 1941			
Pz.Div.	Pz.Rgt.	Pz.38(t)	Bef.(t)
7.	25	166	7
8.	10	118	7
12.	29	109	8
19.	27	116	11
20.	21	116	2

In an analysis of their future needs for 30 operational Panzer-Divisions, on 15 September 1941 OKH determined: *A total of 843 **Pz.Kpfw.38(t)** are needed to outfit six **tschech. Pz.Div.** each with two **Pz.Abt.** with three **le.** and one **m.Kp.** plus three **tschech Pz.Div.** each with two **Pz.Abt.** with two **le.** and one **m.Kp.** There are 386 **Pz.Kpfw.38(t)** (50% of the operational strength on 22Jun41), 56 in the Heereszeugamt Wien, 35 in OKH Reserve Sagan, and 53 already issued to **Pz.Brig.101** for a total of 530 **Pz.Kpfw.38(t)** available to cover this requirement. In addition, another 436 should come out of new production from September 1941 to April 1942 for a grand total of 966, which is an excess of 123.*

7.Panzer-Division with Panzer-Regiment 25

On 11 June 1941, 51 new **Pz.Kpfw.38(t)** were issued to **Panzer-Regiment 25** to fill the three newly formed **leichte Panzer-Kompanien** (one for each **Abteilung**), making it the only **Panzer-Regiment** in the army with 9 **leichte Panzer-Kompanien**. Having started the campaign with 166 **Pz.Kpfw.38(t)** and 7 **Pz.Bef.Wg.(t)**, **Panzer-Regiment 25** was issued 5 **Pz.Kpfw.38(t)** replacements by 18 August and another 5 by 6 September. 124 **Pz.Kpfw.38(t)** and 4 **Pz.Bef. Wg.38(t)** were operational at the start of the second phase of the offensive on 28 September 1941. Having been issued an additional 50 new **Pz.Kpfw.38(t)** by 25 October, and with a relative good replacement parts situation with 92 operational **Pz.Kpfw.38(t)**, the **7.Panzer-Division** was in a somewhat better condition than the other divisions. Still in possession of 131 **Pz.Kpfw.38(t)** and 3 **Bef.38(t)** on 5 December, **Panzer-Regiment 25** reported the following total losses for the period from 6 to 20 December 1941: 51 **Pz.Kpfw.38(t)** and 1 **Pz.Bef.Wg.38(t)** due to mechanical breakdown and another 12 **Pz.Kpfw.38(t)** due to enemy action. The total losses for the entire campaign from 22 June 1941 to 5 January 1942 were reported as: 95 **Pz.38(t)** and 3 **Pz.Bef.** by enemy action, 52 blown up to prevent capture when the front lines were pulled back, 1 captured by the enemy, and 25 returned to Germany with mechanical problems.

7.Panzer-Division Operational Status Reports				
Date	Type	Operational	In Repair	Total Loss
24Jul41	Pz.Kpfw.38(t)	68	65	34
	Pz.Bef.Wg.	7	5	3
23Aug41	Pz.Kpfw.38(t)	43	85	50
	Pz.Bef.Wg.(t)	5	2	2
6Sep41	Pz.Kpfw.38(t)	62	67	59
	Pz.Bef.Wg.(t)	3	3	1
25Oct41	Pz.Kpfw.38(t)	92	50	96
	Pz.Bef.Wg.(t)	4	3	1
2Nov41	Pz.Kpfw.38(t)	88	55	95
	Pz.Bef.Wg.(t)	1	0	0
2Dec41	Pz.Kpfw.38(t)	44	87	108

8.Panzer-Division with Panzer-Regiment 10

The **8.Panzer-Division** under the XXXXVI. A.K. took part in Operation "Marita", the invasion of Jugoslavia, starting on 6 April 1941. After the surrender of Jugoslavia on 20 April 1941, **8.Panzer-Division** reported the total loss of 5 **Pz.Kpfw.II**, 7 **Pz.Kpfw.38(t)**, and 2 **Pz.Kpfw.IV**.

In Heeresgruppe Nord, **Panzer-Regiment 10** started Operation Barbarossa on 22 June 1941 with 118 **Pz.Kpfw.38(t)** and 7 **Pz.Bef.Wg.38(t)**. On 21 November 1941, **8.Panzer-Division** reported: *At this time **Panzer-Regiment 10** has 115 Panzers (out of the 203 with which they started), of which 30 are operational but can't be driven on the slick roads. It is not possible to repair the Panzers in our own **Werkstatt**. They are so worn out after four campaigns that all Panzers need to be overhauled in the assembly plant.*

Above: A Pz.Kpfw.38(t) Ausf.B which has a hull machinegun even though it is assigned to the platoon leader of the 2.Zug/9.Kp./III.Abt./Pz.Rgt.25 in the 7.Pz.Div. prior to the campaign in the east. (MJ)
Below: A Pz.Kpfw.38(t) with the 10.Kp./III.Abt./Pz.Rgt.25 during operation Barbarossa. Pz.Rgt.25 was the only regiment with nine leichte Panzer-Kompanien outfitted with Pz.Kpfw.38(t). (BA 265/37/10)

Above and Left:
As ordered by the Chef H Ruest u. B.d.E. on 30 May 1941:
To increase the Panzers range, every Pz.Kpfw.38(t) and Pz.Bef.Wg.38(t) is to be immediately outfitted with a Betriebstoff-Anhaenger (Fassungsvermoegen 200 l) (fuel trailer with 200 liter drum) and a Ueberpumpvorrichtung (transfer pump rig). (BA 265/28/3A and KHM)

The first replacements sent to the **8.Panzer-Division** were 20 **Pz.Kpfw.38(t)** that were reported to be on the way by rail transport on 11 December 1941.

8.Panzer-Division Operational Status Reports				
Date	Type	Operational	In Repair	Total Loss
22Jul41	Pz.Kpfw.38(t)	58	41	18
	Pz.Bef.Wg.(t)	5	2	0
3Aug41	Pz.Kpfw.38(t)	86	18	14
	Pz.Bef.Wg.(t)	7	0	0
23Aug41	Pz.Kpfw.38(t)	72	27	19
	Pz.Bef.Wg.(t)	7	0	0
4Sep41	Pz.Kpfw.38(t)	78	20	20
	Pz.Bef.Wg.(t)	7	0	0
21Nov41	Pz.Kpfw.38(t)	21		
	Pz.Bef.Wg.(t)	2		
11Dec41	Pz.Kpfw.38(t)	21		
	Pz.Bef.Wg.(t)	3		

12.Panzer-Division with Panzer-Regiment 29

Having started the campaign on 22 June with 117 **Pz.Kpfw.38(t)** and **Bef.Wg.(t)**, there were still 61 operational and 15 total losses one month later on 21 July 1941. They had written off a total of 47 **Pz.Kpfw.38(t)** as complete losses and had been issued 5 replacement **Pz.Kpfw.38(t)** by 18 August. There were still 42 **Pz.Kpfw.38(t)** and 8 **Pz.Bef.Wg.(t)** operational on 21 August.

Panzer-Regiment 29 with their **Pz.Kpfw.38(t)** recorded the following combat action: *On 30 August 1941, at about 1010 hours, the regimental commander sent the **I.Abteilung** with the assignment to set up a defensive position in the area of Popowka train station to shield the right flank of the division. The regimental commander sent a reconnaissance patrol toward the area of Woskressenskoje. The **Abteilung** left at 1030 hours with a strength of 2 Pz.Kpfw.II, 18 Pz.Kpfw.38(t), and 5 Pz.Kpfw.IV.*

The reconnaissance patrol reported the following picture. Russian infantry were located in the area of the Popowka train station. German motorized

Above: A Pz.Kpfw.38(t) Ausf.A in Pz.Rgt.10 with the hole for the hull machinegun covered by an armor plate. Normally Pz.Kpfw.38(t) assigned to platoon leaders and company commanders had the hull machinegun removed to make room for the Fu 5 and Fu 2 radio sets. (MJ)

infantry in the area of the train station reported that a heavy Russian tank had advanced on the road from Tschernikowo toward Hp.

On the basis of the reconnaissance report, the **I.Abteilung** commander sent a company reinforced with two **Pz.Kpfw.IV**s toward the Popowka train station. In cooperation with the motorized infantry located at the train station, the company was given the assignment to advance to the east edge of the village and take up defensive positions. The **leichte Zug** of the **Abteilung** supported by a **Pz.Kpfw.IV** was sent to reconnoiter toward Hp and gain contact with the motorized infantry located there. No enemy activity was encountered.

The opponent in battalion strength holding the area east of the rail line was thrown back toward Tschernikowo having suffered heavy losses. At 1200 hours, four enemy tanks (52 tons) appeared from the south side of Tschernikowo and fired at the **Abteilung** from long range. A short time later, two additional enemy tanks appeared on the rail line about one kilometer northeast of the Popowka train station and charged into the firefight. Our gunfire remained totally ineffective. The opponent destroyed a **Pz.Kpfw.38(t)** and a **Pz.Kpfw.IV**. After the firefight, the tanks pulled back to Tschernikowo. After a short time they again fired at the **Abteilung**. By the concentrated fire of the entire **Abteilung**, the tanks pulled back behind the houses in Tschernikowo. A scouting patrol of the **leichte Zug**, carried out partially on foot, reported that the tanks and elements of Russian infantry were still in the eastern part of Tschernikowo.

Our own losses were one dead, three wounded, a **Pz.Kpfw.IV** totally destroyed, and a **Pz.Kpfw.38(t)** damaged. A heavy Russian 52 ton tank was captured.

The **12.Panzer-Division** lost another 22 **Pz.Kpfw.38(t)** and 2 **Pz.Bef.Wg.38(t)** by 10 November and had only 34 **Pz.Kpfw.38(t)** and 4 **Pz.Bef.Wg.38(t)** operational. They were down to 11 **Pz.Kpfw.38(t)** and 2 **Pz.Bef.Wg.38(t)** operational on 11 December 1941, when 80 **Pz.Kpfw.III** and 20 **Pz.Kpfw.IV** were on the way by rail transport for the **Panzer-Regiment** to be completely reoutfitted with new Panzers. As reported on 18 January 1942, 33 of the surviving 43 **Pz.Kpfw.38(t)** had been sent back home for factory overhaul and 10 had been transferred to **Panzer-Regiment 10**.

19.Panzer-Division with Panzer-Regiment 27

At the start of Operation Barbarossa on 22 June 1941, the **19.Panzer-Division** was reported to have had 116 **Pz.Kpfw.38(t)** and 11 **Pz.Bef.Wg.(t)**.

The following account of **Pz.Kpfw.38(t)** in action on the Eastern Front at the start of the campaign was written on 7 July 1941: The **2.Kompanie/Panzer-Regiment 27** was ordered to attack along the Duna in the direction of Peremerka. The company pulled out after orders had been distributed. The order of march was **1.Zug** (Leutnant Mathieu), **Kompanie-Trupp**, attached **Pz.Kpfw.IV Zug**, **2.Zug**, **3.Zug**. The point quickly reached Myck and made contact with the motorcycle infantry located there. Early in the morning, they had been attacked by superior enemy forces at Point M and had to retreat. Their situation was desperate. Peremerka could not be held.

The company commander of the **2.Kompane/Panzer-Regiment 27** decided to advance with the motorized infantry toward Peremerka and again occupy the cleared area. The **Panzer-Kompanie** started to attack with the same march order. The first enemy fire already occurred on the trail 1 kilometer southeast of Myck. Further penetration was made to the bend in the Duna River. There the **Pz.Kpfw.IV** went into position to provide fire cover. The terrain fell off at a gentle slope, then rose to a height of 800 meters, and a large half circle of forest enclosed the area. There was a lot of brush in the field in between.

Leutnant Mathieu, the commander of the lead platoon, quickly captured the bridge by destroying many machineguns and mortars. He opened the way for the **Panzer-Kompanie** to quickly develop the battle.

12.Panzer-Division Operational Status Reports				
Date	Type	Opera-tional	In Repair	Total Loss
21Jun41	Pz.Kpfw.38(t)	109	0	0
	Pz.Bef.Wg.(t)	8	0	0
5Jul41	Pz.Kpfw.38(t)	86		
	Pz.Bef.Wg.(t)	7		
21Jul41	Pz.Kpfw.38(t)	63*		
21Aug41	Pz.Kpfw.38(t)	46*		
21Sep41	Pz.Kpfw.38(t)	25*		
21Oct41	Pz.Kpfw.38(t)	42		
10Nov41	Pz.Kpfw.38(t)	24	17	68
	Pz.Bef.Wg.(t)	4	2	2
21Nov41	Pz.Kpfw.38(t)	16*		
	*incl.Bef.			

As the point crossed the bridge close behind the bend in the trail, the lead **Pz.Kpfw.38(t)** commanded by Leutnant Mathieu was hit by an anti-tank gun and fell out. Two of the crew were severely wounded. The anti-tank gun was immediately knocked out of action by the following **Pz.Kpfw.38(t)** commanded by Feldwebel Stuppy. The third **Pz.Kpfw.38(t)** was also hit by an anti-tank gun. Unteroffizier Umbach was killed and two of the crew severely wounded. Feldwebel Stuppy identified this anti-tank gun and knocked it out of action. The rest of the Panzers of the lead platoon received heavy fire. They reported the loss of the lead Panzer to the company commander. **Pz.Kpfw.38(t)** tactical number 224 was also hit and caught fire. Unteroffizier Buerstinghaus in this Panzer was severely wounded. While the crew extinguished the fire, a Major from the motorcycle infantry came up and told Gefreiter Markus: "Boys, I'm glad that you came, or we would all have been lost."

As the company commander in his **Pz.Kpfw.38(t)** tactical number 201 turned left to cross the bridge, the Panzer slid sideways into a water-filled hole. Feldwebel Ecken in a **Pz.Kpfw.II** provided effective fire cover as the crew dismounted. The battle was further directed from **Pz.Kpfw.38(t)** tactical number 202. In the meantime the **2.Zug** commanded by Unteroffizier Buerschgens had taken the lead. He drove forward toward the forest, often engaged in action. In the meantime, the remaining elements of the **1.Zug** had entered the battle. Feldwebel Stuppy carried the severely wounded away from the battlefield on his Panzer.

While the **2.Zug** was engaged in a firefight with anti-tank guns and machineguns, the **3.Zug** under Feldwebel Doelcher was employed in defending the flank. Hardly had he taken up position when the enemy in company strength attacked the left flank. Feldwebel Doelcher and his platoon shot up the enemy company. Upon orders from the company commander, all bushes were covered with bursts of machinegun fire. Overall, Russians came out with their hands raised. They were sent to the rear in the hands of the motorcycle infantry. The dismounted crew from the company commander's Panzer had a short firefight with Russians in the area and so took 20 men prisoner.

The dark shadows that appeared in the forest were immediately identified as enemy tanks.

Above: A Pz.Kpfw.38(t) Ausf.D with the 1.Kp./Pz.Rgt.27 of the 19.Pz.Div. on a training exercise in May 1941 prior to the campaign in the East. (NA)

Unteroffiziere Buerschgens, Reppert, and Kowalski shot up four enemy tanks, of which three were 10 to 12 tons and one small swimming tank. In the interim, at close range all platoons had knocked out of action many machinegun nests, heavy 8 cm mortars, and light mortars. The Russian attempts to knock out our Panzers with explosive charges did not succeed.

*The artillery fire increased in intensity. Enemy aircraft also joined in the battle without effect. The **Panzer-Kompanie** did not let itself be disconcerted. They continued to engage ground targets. In the meantime the medium platoon with **Pz.Kpfw.IV**s under the command of Leutnant Giebelthat had also been pulled in and fought exceptionally well.*

*In the interim, the rest of the **Abteilung** came up on the trail through Saborge. The attack itself couldn't be started because orders precluded the unit from entering the forest. Due to the heavy artillery fire, the Panzers were soon ordered by radio to pull back over the bridge. Soon thereafter, we returned back to the rest area.*

*Several attempts during the night and the next evening failed to retrieve **Pz.Kpfw.38(t)** tactical number 201. A new attempt finally succeeded during the night of 10 July. The recovery crew had to survive heavy close combat with the Russians and took an additional 12 prisoners.*

By 4 August 1941, the **19.Panzer-Division** had lost 18 **Pz.Kpfw.38(t)** as total losses due to enemy fire plus one lost on a mine. They were issued 16 **Pz.Kpfw.38(t)** as replacements on 4 October 1941. As reported on 20 December, **Panzer-Regiment 27** had lost all but 50, of which only 10 were operational.

19.Panzer-Division Operational Status Reports				
Date	Type	Operational	In Repair	Total Loss
21Jul41	Pz.Kpfw.38(t)	67	25	24
	Pz.Bef.Wg.(t)	2	8	1
4Aug41	Pz.Kpfw.38(t)	62	37	19
	incl. Bef.38(t)			
25Aug41	Pz.Kpfw.38(t)	57	32	21
	Pz.Bef.Wg.(t)	10	0	1
6Dec41	Pz.Kpfw.38(t)	17	47	68
	Pz.Bef.Wg.(t)	3	3	5
20Dec41	Pz.Kpfw.38(t)	10	40	82

Above and Below Left: Pz.Kpfw.38(t) Ausf.D and E with Pz.Rgt.27 of the 19.Pz.Div. on a training exercise in May 1941. The Ausf.E (left) has blank adapters screwed onto the ends of the M.G.37(t) for automatic fire during the training exercise. The Ausf.E (above) is missing the armor guard on the Nebelkerzenabwurfvorrichtung (smoke grenade rack). (NA)

Right:
A Pz.Kpfw.38(t) Ausf S with the 5.Kp./Pz.Rgt.27 of the 19.Pz.Div. during the last attempt to break through to Moscow late in the Fall of 1941.
(BA)

20.Panzer-Division with Panzer-Regiment 21

Starting the campaign on 22 June 1941 with 116 **Pz.Kpfw.38(t)** and 2 **Pz.Bef.Wg.(t)**, **Panzer-Regiment 21** had written off 27 as total losses by 5 August. In an exchange of "old against new" they had been issued 9 **Pz.Kpfw.38(t)** as replacements by 18 August. On 4 October, another 30 **Pz.Kpfw.38(t)** were issued as replacements to bolster their strength for the second phase of the offensive.

20.Panzer-Division Operational Status Reports				
Date	Type	Operational	In Repair	Total Loss
21Jul41	Pz.Kpfw.38(t)	61	41	16
	incl. Bef.			
5Aug41	Pz.Kpfw.38(t)	45	46	27
	incl. Bef.			
25Aug41	Pz.Kpfw.38(t)	52	46	28
	Pz.Bef.Wg.(t)	2	0	0
6Dec41	Pz.Kpfw.38(t)	45	19	

Panzer Regiment 21 recorded their participation in the final effort to break through to Moscow in the following report dated 6 December 1941:

*After **Infanterie-Regiment 351** attacked out of the Slisnewo bridgehead at 0645 on 1 December, the **III.Abteilung/Panzer-Regiment 21** left the assembly area about 1 kilometer southwest of Pekrowka to attack at 0850, crossed the Nara at the ford or bridge, and reached the Schneisen (demarcation trails in forest) southeast and northeast of Slisnewo. Because the northern Schneise turned out to be impassable, while good progress was made on the southern, the main body of the **Abteilung** struck forward on this trail. On orders from the regimental commander, the reinforced **10.Panzer-Kompanie** advanced to the north on the trail to Datscha, to turn southwest at the crossroad 1 kilometer southwest of Datscha.*

*1Dec41 - The reinforced **1.Panzer-Kompanie** was ordered to support the **Schuetzen** in attacking the long stretch of heights north of Namenskoje and the town itself, especially by suppressing enemy fire expected from the left flank. After occupying the heights, with the attached **1.Panzer-Kompanie**, **Bataillon 112** attacked Namenskoje while the rest of the southern group were shielding the left flank. Because the attack on Namenskoje was progressing well, the reinforced **9.Panzer-Kompanie** was sent east to support **Bataillon 59**. The **9.Panzer-Kompanie** couldn't advance north of Namenskoje because of deep ravines. Therefore they were turned south to support the attack of **Bataillon 112** from the north. After breaking enemy resistance on the north and northwest edge of Namenskoje, the Panzer with **Schuetzen** entered the town, even though the attack of the **schwere Bataillon** from the south had not succeeded. While the town was still being cleared, the reinforced **Panzer-Kompanie** was sent to attack Klowo. After crossing the stream in the village, they struck forward on both side of the trail toward Klowo. About 300 meters from the east edge of Klowo, two **Pz.Kpfw.38(t)** ran onto a minefield that couldn't be bypassed. The mine clearing troops from the **Bataillon**, which were immediately requested by radio, didn't arrive until after dark, just like both **Schuetzen-Kompanie** from the **schwere Bataillon**.*

*In the interim, the division operations officer passed on the order to the commander of the **III. Abteilung** that all elements except the reinforced **1.Panzer-Kompanie** were to win through to **Panzer-Regiment 21** in Datscha in order to still take Matschichino in the evening.*

*After succeeding in gapping the minefield, as ordered the reinforced **1.Kompanie** with the **2.Kompanie** of the **schwere Bataillon** started to attack toward Klowo, which, based on a report from **Schuetzen-Regiment 59**, had already been taken by **Btl.59**. As they neared the town, the attacking **Kompanie** received heavy rifle and machinegun fire from the southern part of the town. Also, the town was still occupied by the enemy. The **Schuetzen** explained that an attack after nightfall was impossible and returned back to Namenskoje with the reinforced **1.Kompanie**. **Schuetzen-Regiment 59** ordered the **schwere Bataillon** together with the reinforced **1.Kompanie** to clear the wood lines south of Klowo at dawn.*

*2Dec41 - After completing this assignment, the reinforced **1.Panzer-Kompanie** again supported the **Schuetzen** attack on Klowo, because these received heavy fire from the forest southeast of Klowo. Together with the **Schuetzen**, the reinforced **1.Panzer-Kompanie** occupied Klowo and then defended against the forest to the southeast. Then about noon they received orders to support the further advance of **Bataillon 112** that had advanced farther to the northeast. Because the trail to it and the edge of the forest northeast of Klowo were mined, the **Kompanie-Chef** asked permission from **Schuetzen-Regiment 69** to detour by way of Datscha to follow the Regiment or be given **Minensucher** (mine clearers). Both requests were denied. While carrying* out their order, the **Kompanie-Chef (Pz.Kpfw.IV)** and a **Zugfuehrer (Pz.Kpfw.38(t))** drove onto mines. Due to mines and artillery hits as well as mechanical breakdown, the reinforced **1.Panzer-Kompanie** had shrunk to three Panzer. They then now received orders from **Schuetzen-Regiment 59** to follow the Regiment to Datscha. One Panzer, which didn't receive the radio message to leave Klowo, remained there and took part in repulsing Russian attacks on Klowo. After repairing additional vehicles during the night, the rest of the **1.Panzer-Kompanie** struck through to the Regiment in Matschichino during the morning of 3 December. All of the officers in the **1.Kompanie** were incapacitated.

In the interim, on 1 December, the **Stab III. Abt**. with the **9.Panzer-Kompanie** and elements of the **12.Panzer-Kompanie** left Namenskoje and reached Slisnewo at dusk. From there, guided by an officer from **Infanterie-Regiment 351** who supposedly knew the local area, they started to advance toward Datscha. Because this officer lost his way in the forest several times, the **III.Abteilung** arrived in Datscha about midnight. Together with two battalions from **Infanterie-Regiment 351**, the **III.Abteilung** leaguered in the town, repulsed two Russian attacks during the morning of 2 December, and supported the advance of **Infanterie-Regiment 351** on Mogutowo with two reinforced Panzer platoons. At 1330, the remaining elements of the Regiment arrived in Datscha.

On 1 December about 1100, the left **Kampfgruppe**, consisting of the reinforced **10.Panzer-Kompanie**, **Pionier-Zug/III.Abteilung**, **Stab I./Artillerie-Regiment 92**, **Infanteriegeschuetz-Zug/Bataillon 59**, and **Pak-Zug/2.Kompanie/Panzerjaeger-Abteilung 92**, started on the trail to Datscha. Because of uncertain map detail and the forest preventing observation, time-robbing trail scouting was necessary. The road fork that should have turned south couldn't be found in spite of long searches. About 1330 the **Kampgruppe** met the right **Bataillon** of **Infanterie-Regiment 351**, which after it had overcome heavy resistance had also started moving east. At 1500 the division ordered: "Advance to the northeast and occupy

Right:
Two Pz.Kpfw.38(t) from Panzer-Regiment 21 of the 20.Panzer-Division are attempting to recover a third that has broken through the ice.
(HLD)

Matschichino. **Schuetzen** and the main body of Panzer will follow." The advance was significantly delayed by several strong tree obstacles that were mined. After five **Pz.Kpfw.38(t)** and two **Pz.Kpfw.IV** had been lost on mines, the Regiment ordered the advance to be halted and leaguer for the night where the trails crossed 1-1/2 kilometers north of Datscha. About 2000, the Regiment received orders by radio to take Matschichino with the attached **II.Bataillon/Infanterie-Regiment 351** and **I.Abteilung/Artillerie-Regiment 92**.

2Dec41 - Because of problems observing the layout of the terrain and lack of signal contact, the **Kampfgruppe** didn't join up with the **Bataillon**. Reconnaissance during the night and at dawn reported obstacles on the trail to Matschichino with strong enemy defenders. At the same time strong pressure was felt from the enemy returning here from the southwest. During the morning this was broken up and they moved off to the north.

About 1130, advancing from Datscha toward Mogutowo, the **I.Bataillon/Infanter-Regiment 351** met the **Kampfgruppe**. Up to 1330, the **Panzer-Kompanien** (without the **1.Kompanie** in Datscha) took up positions with the **I.Abteilung/Artillerie-Regiment 92** just north of Datscha. At 1420, with the attached **MTW-Kompanie**, a **Pionier-Zug/2.Kompanie/Panzer-Pionier-Bataillon 92** and a **Pak-Zug/2.Kompanie/Panzer-Jaeger-Abteilung 92**, as well as the **I.Abteilung/Artillerie-Regiment 92**, the Regiment detoured south to attack Matschichino, met **Bataillon 112** advancing from the southwest at 1515, and together with them took the town as night fell. The enemy put up only weak resistance and escaped to the east with few losses.

3Dec41 - During the night and throughout the next day, the Regiment defended Matschichino with **Bataillon 112** and a **leichte Flak-Batterie** that had been sent. Combat reconnaissance by Panzers in the bordering woods toward Staroswetino and Kuwekino reported weak enemy forces in the first village and strong enemy forces in Schubino; otherwise no enemy activity. From this reconnaissance it was proven that at this position the offensive had penetrated 10 kilometers deep into the enemy position.

4Dec41 - In compliance with the directive received the previous evening from the division, after setting it on fire, the Regiment cleared out Matschichino at 0715 and started the return trip through Datscha to Slisnewo. Also providing the rear guard for **Infanterie-Regiment 351** retiring from Mogutowo, with its last elements the Regiment reached Slisnewo at 1030 and settled in Balabanowo.

Influence of Weather on the Battle

Slick roads caused travel times of vehicles with and without cleats to be significantly lengthened. Small hills and slopes decreased the speed even with cleats installed. Rapid movement in combat was greatly hindered. The number of places in which terrain could be crossed by Panzers was greatly reduced, resulting in the Panzers bunching up. The probability that obstacles could be pushed over was very low.

It was necessary to start the vehicles at least every two hours. This disturbed the troops' sleep and increased fuel consumption. Assembly areas had to be selected farther to the rear. Thinner oil was needed in the low temperatures.

Attack in snowstorms or even in powdery snow resulted in obstructing vision from the sights and vision ports. Driving in snow even when it isn't falling is especially stressful for the driver because his vision is obstructed. Turrets must be traversed now and then to keep from being frozen stuck. Snow glasses are needed for the driver and commander. Crossing fords can lead to suspension parts being frozen stuck. A high number of Panzers even with cleats got stuck in snow cover over 20 cm deep because the depth of drifts and holes can't be recognized.

Repairs in low temperatures and snow are extremely difficult. Broken springs and suspension damage of all types occur on hard frozen and packed trails. Towing broken-down vehicles is more difficult on slick roads. The **Zugmaschinen** absolutely need snow chains.

Kalkanstrich (whitewash) is absolutely necessary on vehicles and must be adapted to the current snow conditions.

Impression of the Enemy, Mines, and Bunkers

Never encountered before in such numbers, mines were very cleverly camouflaged and rarely spotted under the snow, leading in every case to immediate loss of the vehicle. Driving onto a mine resulted in torn hulls and significant suspension damage for about 50% of the **Pz.Kpfw.38(t)**. **Pz.Kpfw.IV** usually had suspension damage. In all cases, drivers or radio operators in **Pz.Kpfw.38(t)** were wounded but in **Pz.Kpfw.IV** were usually only lightly injured.

Individual tank traps were extraordinarily cleverly camouflaged but usually didn't cause damage to the Panzers because the strong covers let the Panzer break through slowly and there weren't any mines at

Above: This Pz.Bef.Wg.38(t) Ausf.E or F (with a Pz.Kpfw.38(t) Ausf.B - D Turm) belonged to the Panzer-Nachrichten Abteilung in the 20.Panzer-Division. (KHM)
Below: A Pz.Kpfw.38(t) Ausf.E and an Ausf.D with Panzer-Regiment 21 of the 20.Panzer-Division. (MJ)

the bottom of the trap. After significant work digging out earth, one Panzer could drive out under its own power.

Trees were downed for obstacles, sometimes for 50 meters deep and mined all round. The electrical mine detectors didn't respond since the mines were laid in wood boxes. Poking wasn't successful because of the frozen ground. It is thought that mined locations are posted by notching trees around the area.

Large numbers of bunkers were cleverly located in deep woods and strongly manned. These bunkers, usually located off to the side of the trails, are very difficult to knock out by Panzers, which almost always are restricted to the trails.

Again it was proven that Panzers have limited use in forested terrain.

Panzers serve primarily to bolster the morale of the **Schuetzen**. The demand that Panzers charge into mined positions should be declined, because this only leads to loss of this valuable equipment. Understanding of this is still partially lacking.

The enemy fought bravely in holes and bunkers. Overall, it was observed that his positions were heavily occupied. His winter uniform (fur caps and leggings) were good and usable. The field of fire for bunkers was limited in favor of camouflage. He didn't have many heavy weapons and artillery. Only a small number of handheld weapons and machineguns were captured. One anti-tank gun and a tank were knocked out.

<u>Supply</u> - Resupply wasn't especially difficult. However, it is necessary on slick roads to leave several **Pz.Kpfw.II** with the **Kampfstaffel B** in order to bring supply vehicles forward.

Our Losses: Three killed, 13 wounded, 110 light frostbite (all remained in service), 68 sick (all remained in service), 0 missing.

Losses of Own Weapons - As a result of pulling back the front, 5 **Pz.Kpfw.38(t)** and two **Pz.Kpfw.IV** were left behind in enemy territory after valuable parts were removed and they were blown up.

On 4 December 1941, **Panzer-Regiment 21** reported on attempts to recover Panzers, as follows:

The following **Panzerkampfwagen** were lost east of the Nara River during the battles on 1 to 3 December 1941

1. *Pz.Kpfw.38(t)* tactical number 16 ran onto a mine. Both tracks broken, hull fractured, and left drive wheel torn off. Total loss. Two wounded.

2. *Pz.Kpfw.38(t)* tactical number 24 (**Fgst.Nr.344**) direct hit by anti-tank mine penetrated rear armor and engine. Total loss.

3. *Pz.Kpfw.38(t)* tactical number 15 (**Fgst.Nr.756**) ran onto a mine. Left hull fractured, all roadwheels deformed, radio and internal equipment destroyed. Total loss. Two wounded.

4. *Pz.Kpfw.38(t)* tactical number 2 (**Fgst.Nr.832**) ran onto two mines. Entire rear torn open and fractured, and roadwheels deformed. While being recovered it hit another mine, resulting in the front part of the hull being torn open. Total loss. Two wounded.

5. *Pz.Kpfw.38(t)* tactical number 13 (**Fgst.Nr.910**) ran onto a mine. Forward left hull side torn wide open and left roadwheels deformed. During recovery it hit another mine and the right hull side and right roadwheels torn off. Total loss. Two wounded.

6. *Pz.Kpfw.IV* tactical number 69 mechanical breakdown - two springs suspension arm broken, not drivable.

7. *Pz.Kpfw.IV* tactical number 75 (**Fgst.Nr.80848**) mechanical breakdown - side drive shaft broken. Conditionally drivable.

8. *Pz.Kpfw.IV* tactical number 65 (**Fgst.Nr.82117**) ran onto a mine. Left track broken, left drive wheel bent and broken, forward left spring retainer and roadwheel deformed. Not drivable.

9. *Pz.Kpfw.38(t)* tactical number 31 ran onto a mine. Right rear hull side torn open, track and rear roadwheels deformed. Total loss. Two wounded. Towed back to Slisnewo with a **Zugmaschine** on 2Dec41.

10. *Pz.Kpfw.38(t)* tactical number 32 ran onto a mine. Track and front left roadwheel torn off. Not drivable. Towed back to Slisnewo with a **Zugmaschine** on 2Dec41.

11. *Pz.Kpfw.IV* tactical number 60 ran onto a mine. Right track broken, right front spring retainer and roadwheel deformed. Not drivable.

12. *Pz.Kpfw.IV* tactical number 63 ran onto a mine. Left drive wheel torn off and track broken. Not drivable.

13. *Pz.Kpfw.38(t)* tactical number 27 ran onto a mine. Track broken and forward right roadwheel deformed. Not drivable. Towed back to Slisnewo by two *Pz.Kpfw.38(t)* on 3Dec41, it couldn't be towed by a **Zugmaschine** because of road conditions.

14. *Pz.Kpfw.38(t)* tactical number 5 drove into a 3 meter deep anti-tank ditch. No damage. **Panzerpionieren** used explosives to destroy the wall, and after it was dug out, sent back to the troops under its own power.

15. *Pz.Kpfw.IV* tactical number 51 mechanical breakdown - forward left spring retainer broken.

Conditionally drivable.

On the evening of 3 December, two Pz.Kpfw.38(t) (listed as 1. and 2.) were towed back in the valley on the southern trail to Slisnewo. A Pz.Kpfw.IV (listed as 6.) was driven there under its own power and the final drive completely failed while attempting to climb up a steep slope. Because of the slick roads and the steep slope, it was impossible to tow away these three Panzers. After removing the machineguns, breech blocks, radio sets, batteries, sights, jacks, track, roadwheels, spring bundles, fuel pumps, and all other parts that could be removed in the short time available, these Panzers were blown up or set on fire so that they wouldn't fall intact into enemy hands. They were blown up early on 4 December, because members of the Werkstattkompanie were warned that it was high time they pulled back, since the enemy was closely following.

Pz.Kpfw.38(t) (listed as 3., 4., and 5.) were recovered from the large minefield west of Klowo and towed by Zugmaschinen to Namenskoje, where two of them ran onto mines again. At Namenskoje the ice-covered slope onto the bridge was not passable with Panzers in tow. The Zugmaschinen themselves could get up this slope only by using their winches. All three Panzers were blown up after usable parts and equipment were removed.

4.(Pz.) Kp./Fuehrer-Begleit-Bataillon

As ordered by the OKH Chef H Ruest und BdE on 18 June 1941: *The **Panzertruppenschule** is to immediately create a new **Pz.Kpfw.Kp.-Fuehr.Begl. Btl.** from its **Lehr-Kompanie** by 19 June 1941. Sixteen fully equipped **Pz.Kpfw.38(t)** with radio sets, equipment and tools are to be acquired by the **Pz.Tr.Schule Wuensdorf** at H.Za.Magdeburg on 17 June 1941 and driven by road to Wuensdorf.*

As recorded in the war diary for **Kampfgruppe Naehring**: *23Sep41- On orders from Hitler, a **Kampfgruppe** was created from the **4.(Pz.)Kompanie** and elements from the **2. and 3. Kompanie./Fuehrer-Begleit-Bataillon** that is to be employed by Heeresgruppe Nord in the Leningrad area. Hptm. Naehring was ordered to be the **Kampfgruppe** commander.*

Assigned to A.O.K.18 on 27 September 1941, the 16 **Pz.Kpfw.38(t)** with the **4.(Pz.) Kp.** arrived in Narwa on 29 September. Preparations to attack the Oranienbaumer Kessel, were abandoned when a scouting trip on 24 October revealed that the terrain was impassable for Panzers except on the roads. After a period of spent in training, the **4.(Pz.)Kp.** was then transferred to A.O.K.16 and attached to the **126.Inf. Div.** on 12 November 1941.

During the period from 13 November until 9 December 1941, the **4.Panzer-Kompanie/Fuehrer-Begleit-Bataillon** was sent into action to support the infantry of the **126.Inf.Div.** in combat usually against numerically superior opponents at Krassn.Wischerka, Nekrassowo, Aleksandrowskskoje, Bol.Wischera, and Weretje. During this period three men had been killed, 18 wounded, and five **Pz.Kpfw.38(t)** written off as total losses due to enemy action. (one hit a mine, one from a direct hit by artillery, and one from a heavy mortar destroying the suspension; two were recovered and returned to Germany for major overhaul). During 18 days in combat they never reported encountering a single enemy tank and fired 788 rounds from the **3.7 cm Kw.K.** and 114,200 machinegun rounds. On 19 November, strafing aircraft had repeatedly attacked the Panzers but the 2 cm shells ricocheted off without causing any damage.

On 7 December 1941, the **4.Pz.Kp.** was ordered to transfer their remaining **Pz.Kpfw.38(t)** to the **8.Pz.Div.** Then on 15 December, A.O.K.16 ordered that they were to get their **Pz.Kpfw.38(t)** back from the **8.Pz.Div.** Five were brought back on 16 December and another four by 21 December 1941.

The **4.(Pz.)Kp.** was sent back into action with Kampfgruppe Naehring on 20 December 1941 to guard the Grusino bridgehead. Two **Pz.Kpfw.38(t)** rescued an ambushed supply column and another two sent on a reconnaissance mission set a village on fire with HE shells causing about 60 enemy soldiers to abandon the village. The **Kompanie** with eight operational **Pz.Kpfw.38(t)** was ordered to support the **126.Inf.Div.** on 21 December 1941. Losing one Panzer damaged by a mine, the **2.Zug** with three **Pz.Kpfw.38(t)** broke through and rescued elements of the division, including **Inf.Rgt.424**, which had been cut off by a strong enemy force.

In action again on 28 December 1941 with six operational **Pz.Kpfw.38(t)** to attack a Russian breakthrough, the **Kompanie** was ordered to send the last three damaged Panzers to Tschudowo for repair on 31 December 1941. The **4.(Pz.)Kp./Fuehrer-Begl.Btl.** reported a strength of eight **Pz.Kpfw.38(t)** from 1 January to 21 February 1942, seven on 1 March, and six on 11 and 21 March 1942.

Campaign in the East 1942/43

On 29 January 1942, the Chef der Heeresruestung und Befehlshaber des Ersatzheeres sent the Chef der Generalstabes Generaloberst Halder a memo on the equipment situation for 1942 including the **Panzerlage** (tank situation) for the **Pz.Kpfw.38(t)**: *After deducting the number needed for planned issue to the **22.Pz.Div.**, aid to Hungary, and 40 Pz.Kpfw.38(t) for Panzerreserve Sagan, there will be only 86 Pz.Kpfw.38(t) from new production available by the end of April 1942 for rebuilding the Panzertruppen in the Feldheer.*

*Rebuilding can be accomplished by Case A: equal distribution to all **Panzertruppen** in the **Feldheer**, or Case B: concentration to create an **Operationsarmee**. In Case A, 426 Pz.Kpfw.38(t) are needed to fill the six kl.Pz.Div. (each with two Pz.Abt. with two le.Pz.Kp. and one m.Pz.Kp.) in the Ostheer. Situation reports gathered on 22 December 1941 reported 159 Pz.Kpfw.38(t) operational and 136 in repair by units. Another 69 Pz.Kpfw.38(t) were sent as replacements during the period from 3 to 31 December 1941.*

*In Case B, one **Panzer-Abteilung** with 35 Pz.Kpfw.38(t) is needed to help create a third Panzer-Abteilung for the Panzer-Divisions in Heeresgruppe Sued and to create a new Panzer-Abteilung for one of the five Infanterie-Division (mot). This would leave only 86-35 = 51 Pz.Kpfw.38(t) for rebuilding the Panzer-Divisions in the rest of the Ostheer. Therefore, the 10 Panzer-Divisions (including the 6., 7., 8., 19., and 20.Pz.Div. with Czech Panzers) remaining in Heeresgruppe Mitte and Nord will only have sufficient Panzers available for one Panzer-Abteilung each. <u>In accordance with Hitler's directive, these 10 Panzer-Divisions don't need to be strategically operational but only need to be suitable for eliminating local problems and can be outfitted with rebuilt Panzers.</u>*

On 18 February 1942, Gen.St.d.H. Org.Abt.(I) issued orders for rebuilding the **Ostheer** in the Spring of 1942. *Eight to 10 Panzer-Abteilung are to be taken from the Panzer-Divisions in Heeresgruppe Mitte and used to create a III.Panzer-Abteilung for Panzer-Regiments in Heeresgruppe Sued or to create Panzer-Abteilungen for the motorized Infanterie-Divisions. It is intended to transfer the 6., 7., and 20.Panzer-Divisions from Heeresgruppe Mitte to the West for rebuilding..*

The remaining units in Heeresgruppe Mitte and Nord that haven't been fully rebuilt are to concentrate their available Panzers to create mobile Kampfgruppen. With the exception of the 1., 3., and 5.Panzer-Divisions, the rest of the Panzer-Divisions in Heeresgruppe Mitte and Nord are to be reorganized with one Panzer-Abteilung with two or three le.Pz.Kp. and one m.Pz.Kp.

With the **6.** and **7.Panzer-Division** transferred back to the West and the **12.Panzer-Division** converted to **Pz.Kpfw.III**, only three Panzer-Divisions (the **8.**, **19.** and **20.**) outfitted with **Pz.Kpfw.38(t)** were left in the East. In addition, when the **7.Panzer-Division** was sent back to the West, its remaining **Pz.Kpfw.38(t)** were transferred to the **1.** and **2.Panzer-Division**. Thus, with the exception of the new **22.Panzer-Division** sent to Heeresgruppe Sued, the rest were Panzer-Divisions in name only and left to contend with local emergencies usually with less than a full **Panzer-Abteilung**.

1.Panzer-Division with II.Abt./Pz.Rgt.1

Kompanie Brahm with 12 **Pz.Kpfw.38(t)** from **Panzer-Regiment 25** attached to **Panzer-Regiment 1** on 20 February 1942 had 5 operational and had 4 total losses from mines by 23 February. As the **4.Kp./Pz.Rgt.1** it still had 7 **Pz.Kpfw.38(t)** on 27 March, having lost five.

Panzer losses by **Kampfgruppe Koll** with the **1.Panzer-Division** were reported on 4 April 1942:

The extremely bad snow conditions prevent almost every unaided movement of the Panzers other than on shoveled-out paths. Scouting paths and shoveling can't be done under fire and are limited by a shortage of manpower. All available manpower must be employed to build a supply road. Well-directed artillery fire can achieve direct hits on Panzers stuck in open terrain.

Panzers are knocked out by T 34 at ranges of 200 to 800 meters. The Panzer 38 (t) can't destroy or repulse a T 34 at these ranges. Because of its gun, the T 34 can knock out an attacking Panzer at long range.

Because of the frequent changes to the HKL (main battle line), it wasn't always possible for Panzers to disengage from the enemy or be recovered in time. The situation doesn't allow infantry to guard a Panzer in front of the HKL. Three Panzers had to be blown up.

Total losses during the period from 30 March to 3 April 1942 amounted to:
On 31 March 1943 in the 3.Kompanie Koenig:
2 Pz.38(t) - on the road from Adrimaja to Medwediza (300 m west of Medwediza), both with transmission

failure and hits in the suspension. Located directly in front of the enemy lines. Attempt at recovery failed on 3 April 1942.

1 Pz.38(t) - hit by T 34 in crew compartment and engine, blown up.

3 Pz.38(t) - stuck in snow and had to be blown up when the HKL was pulled back. Towing with a Pz.IV wasn't successful. Located west of Medwediza within the enemy lines.

1 Pz.38(t) - direct hit by artillery.

1 Pz.IV - hit in turret by T 34 and burned out. Located west of Medwediza.

On 1 April 1942:

1 Pz.38(t) - hit by T 34 and crew compartment totally destroyed. Located northeast edge of Jljmowka.

1 Pz.38(t) - hit by T 34 and burnt out. Located on west edge of Jljmowka.

1 Pz.IV - ran onto a mine, turret displaced and burned out. Located 200 meters west of Nachrotovka.

The tactical employment of the Panzers was correct. Scouting the terrain. Providing artillery covering fire for every attack by the Panzers with all available batteries (two heavy and three light). Closest cooperation between Panzers and infantry.

Our own Infanterie won't hold against tanks. The appearance of an enemy tank brings the attack to a halt and in several cases to immediate rearward movement, resulting in our own Panzer standing alone far in front of our own lines and soon becoming stuck, and were then the targets of enemy artillery, tanks, and anti-tank guns.

On 1 April, two Pz.38(t) with transmission failure and one Pz.38(t) hit in the turret by an anti-tank gun, and on 3 April, one Pz.38(t) with differential failure were towed to Oljenin for repair. Transmissions were repaired in two Pz.38(t), which were returned to the troops on 3 and 4 April 1942. The time it will take to repair the other two Panzers is unknown. Repair parts were requested by radio from Pz.Rgt.25 on 2 April.

Panzer status at noon on 4 April 1942: 14 Pz.38(t), two of them in Oljenin, three Pz.IV, one of which is being repaired.

On 23 April 1942 there were 21 **Pz.Kpfw.38(t)** in the **3.Kp./Pz.Abt.Herschel (7.Kp./II.Abt./Pz.Rgt.1)** that had been taken over from **Pz.Abt.Schroeder.** 14 were operational on 13 June, 18 on 27 June, 6 on 9 July, 10 on 15 July which had been distributed as two or three each in the **Stab, 2.Kp., 7.Kp., and 8.Kp./II./Pz.Rgt.1.** The **1.Pz.Div.** still had 19 **Pz.Kpfw.38(t),** of which 16 were operational on 1 August, 5 on 15 August, and 8 on 18 November. The remaining three **Pz.Kpfw.38(t) Fgst.Nr.1102, 1302, 1303** were transferred to the **20.Pz.Div.** on 20 December 1942, when the **1.Pz.Div.** was transferred back to the West.

2.Panzer-Division with II.Abt./Pz.Rgt.3

33 **Pz.Kpfw.38(t),** of which 17 were operational, were taken over from the **7.Panzer-Division** on 20 June 1942. They still had 28 **Pz.Kpfw.38(t),** of which 24 were operational on 15 July and 26 on 5 August. But they were down to 6 operational on 15 August, and had 7 total losses in action during the period from 5 to 23 August. The **II.Abt./Pz.Rgt.3** received five new **Pz.Kpfw.III lang** and turned 11 **Pz.Kpfw.38(t)** back in on 23 August 1942.

7.Panzer-Division with Panzer-Regiment 25

Having total losses of 175 out of 238 **Pz.Kpfw.38(t)** and 11 out of 16 **Pz.Bef.Wg.**, the **7.Panzer-Division** was left with 63 **Pz.Kpfw.38(t),** of which only 8 were operational, and 5 **Pz.Bef.Wg.** on 10 January 1942. 16 were operational on 12 February and 27 on 19 February, with a total inventory of 63 **Pz.Kpfw.38(t)** and 1 **Pz.Bef.Wg.38(t)** still reported to be available on 5 March 1942. A total inventory of 74 **Pz.Kpfw.38(t)** and 1 **Pz.Bef.Wg.38(t)** was reported on 12 March 1942, of which 40 were in the **Werkstatt Pretschistoje.** Of these, 22 were assigned to **Pz.Abt. Schlothaue (II.Abt./Pz.Rgt.25** with the **5., 6., 7.,** and **8.Kp.),** 12 to **Pz.Kp.Reinhardt**, and 40 to **Pz.Abt. Schultz (I.Abt./Pz.Rgt.25** with the **1., 2.** and **4.Kp.).**

8.Panzer-Division with Panzer-Regiment 10

Panzer-Regiment 10 still had 34 **Pz.Kpfw. 38(t),** of which 9 were operational on 20 January 1942. They had lost 12 and were down to 22, of which 12 were operational on 5 March. The **8. Panzer-Division** was issued 25 **Pz.Kpfw.38(t)** on 7 March which arrived on 13 March and were used to outfit **Pz.Kp.Motsch** giving the **Panzer-Regiment** a total of 47 **Pz.Kpfw.38(t)** at the end of March, of which 47 were operational.

Another 20 **Pz.Kpfw.38(t)** replacements arrived for the **8.Pz.Div.** on 18 May. A total of 65 **Pz.Kpfw.38(t)** were available on 28 June 1942, of which 25 were operational in the **I.** and **III.Pz.Abt./Pz.Rgt.10** with the **1., 2., 3., 4., 9.,** and **12.Pz.Kp.** Only 8 to 11 were operational in August through mid-September but the number had increased to 17 by 30 September, down to 15 on 12 October, 14 on 15 November, 29 on 30

November, and 16 by 13 December. They still had 42 **Pz.Kpfw.38(t)** on 1 January 1943, of which 24 were operational with the **I.Abt./Pz.Rgt.10** with 3 **le.Pz.Kp.** and 1 **m.Pz.Kp.**

The **I.Abt./Pz.-Rgt.10** reported the following total losses in action on 7 and 11 January 1943:

Unit	Fg.Nr.	Damage	Disposal
1.Kp.	1483	Hit by AT gun	Total writeoff
1.Kp.	1173	Blown up	Total writeoff
1.Kp.	1179	Gun tube burst	Riga
2.Kp.	1263	Blown up	Total writeoff
2.Kp.	782		Total writeoff
3.Kp.	1291	Hit by AT gun	Total writeoff
3.Kp.	504	Hit by AT gun	Total writeoff
3.Kp.	205	Hit by AT gun	Total writeoff
3.Kp.	265	Hit by AT gun	Total writeoff
2.Kp.	1169	Blown up	Total writeoff
2.Kp.	1261	Mines	Riga.

This left 10 out of 34 operational on 10 January 1943. They were up to 36 **Pz.Kpfw.38(t)**, of which 29 were operational on 7 February, but had been reduced to 27, of which 12 were operational by 19 February.

19.Panzer-Division

Left with the **I.Pz.Abt./Pz.Rgt.27** with the **1., 2., 4., and 5.Kp.**, the unit still had 34 **Pz.Kpfw.38(t)** on 15 July 1942. There were 2 total losses reported on 21 July, but 10 newly issued **Pz.Kpfw.38(t)** brought their strength back up to 42 on 28 July. There were 3 total losses in August, and 4 new replacements were reported on 2 September. There was another total loss reported on 7 October, and 22 **Pz.Kpfw.38(t)** were sent back to the Heereszeugamt Magdeburg on 26 November 1942.

19.Panzer-Division
Pz.Kpfw.38(t) Operational Status Reports

Date	Total Available	Operational	In Repair	Total Loss
15Jul42	34	26	8	
28Jul42	42	37	5	2
12Aug42	41	41	0	
26Aug42	39	30	9	2
16Sep42	42	42	0	
30Sep42	42	41	1	
14Oct42	41	39	2	1
28Oct42	41	32	9	
11Nov42	41	38	3	
26Nov42	19	19	0	

20.Panzer-Division with III.Abt./Pz.Rgt.21

Combat report from the **9.Kompanie/Panzer-Regiment 21** during the period from 5 to 21 January 1942 while employed with the XX.Armee Korps

5Jan42 - On orders from the **20.Pz.Div.**, the **9.Kp./Pz.Rgt.21** with 10 **Pz.38(t)** led by Lt. Dreher was sent to Wereja. There the **Kompanie** was attached to **Gruppe Major Seidensticker (5.Pz.Div.)**. Objective: Knock back the opponent that had broken through the XX A.K. front.

6Jan42 - The opponent pushed forward toward the northwest over the Rut sector in order to gain control of the Wereja to Medyn road. The situation demanded immediate Panzer action. One platoon was sent to attack Blayoweschtschenskoje; another platoon was sent to attack Makarowo. The enemy were cleaned out of both villages by the advance of our own Panzers. Our own forces again occupied these villages. Two Panzers were left behind in the village to guard Blayoweschtschenskoje.

7Jan42 - During the night the strong enemy forces managed to advance to the north in the Prowma valley and occupy Wyschgorod. In the morning, the Panzers attacked and took Wyschgorod. The deep snow restricted the advance to single file on the road and thereby made the attack more difficult. During the following night the Panzers were left to guard Wyschgorod.

8Jan42 - At dawn the rest of the **Panzer-Kompanie** was pulled out of Wyschgorod and placed at the ready for use by Btl. Bleckwend in Nowo Borisowo. Soon afterward the enemy again managed to retake Wyschgorod. A counterattack with infantry was broken off because of the heavy infantry losses.

Together with the infantry the Panzers left to guard Blayoweschtschenskoje successfully repulsed a strong attack from the south. Lt. Dreher was wounded by a shell fragment, and Lt. Winter took over command of the **Kompanie**.

9Jan42 - With artillery and rocket fire the opponent attempted to attack Blayoweschtschenksoje, Polmeschajewo, and Nowo Borisowo along the entire sector in a northwesterly direction in order to reach the Wereja-Medyn road. The attack on Blayoweschtschenskoje was knocked back with support from the Panzers.

In the center the opponent entered Teremeschajewo but was soon thrown back by a counterattack by four Panzers from the **10.Kompanie**. The enemy attack on Nowo Borisowo was repulsed by *the guarding Panzers and **Infanterie**.*

*10Jan42 - Guarding Blayoweschtschenksoje and Nowo Borisowo without heavy enemy pressure. Because the opponent pressed forward in the Prowma valley, part of the **10.Kompanie** was defending Panowo toward the east. During the night, strong enemy groups attacked Cottjajewo but were knocked back by the help of Panzers from the **10.Kompanie**.*

*11Jan42 - Strong enemy pressure toward the northwest on Cottjajewo and Serenskoje. The opponent was thrown back from Serenskoje in a counterattack supported by two Panzers from the **10.Kompanie**. In the center, an enemy recon troop was destroyed by Panzers on guard southeast of Duchkino.*

*12Jan42 - The **9.Kompanie** is scattered at one to two Panzers in each village in a sector over 14 kilometers wide. At dawn the enemy pushed strong forces toward Jegorje with the intention of advancing to the north on the Wereja-Medyn road. The attack was knocked back with the support of two Panzers.*

*13Jan42 - The Panzer on guard near Dudkino was relieved by Kampfgruppe Chevallerie. Near Nowo Borisowo the opponent managed to bypass and surround the **Kampfgruppe** located there. In a retreat to Gorki, Btl. Bleckwend used both Panzers assigned to them as a **Nachspitze** (rear guard). Later as ordered by the battalion commander, both Panzers had to be blown up because they had both broken down. In the evening, Oblt. Dittmer took over the **Kompanie** and on orders from Major Seidensticker assembled the **Kompanie** in Wasilewo.*

*14Jan42 - Enemy pressure increased in the northerly direction. The opponent with about two battalions attempted to strike through the woods between Panowo and Sotnikowo. The **Kompanie** was sent to attack the woods and, causing heavy losses, knocked the enemy back toward the south. During the night the Kompanie guarded the trail from Sotnikowo to Panowo.*

*15Jan to 18Jan42 - With two attached s.**M.G.- Gruppen** the **Kompanie** guarded the trail from Panowo to Sotnikowo. In spite of recurring enemy attacks supported by strong artillery preparations and rocket fire, he didn't manage to advance farther north.*

*18Jan42 - As ordered by Major Seidensticker, the **Kompanie** was pulled out and started the return march by way of Archangelskoje, Troparewo, Autobahnkreuz, Autobahn to the west.*

*21Jan43 - **9.Kompanie** arrived in Rjabzowa.*

Losses: Personnel: two killed, 15 wounded, and 2 missing. On 8 January, after the Panzers pulled back, the opponent surprisingly pushed into Wyschgorod. One Panzer rolled over, and after recovery wasn't possible because of the snow, the crew attempted to destroy the Panzer. During this attempt the crew was overrun by the enemy.

*Equipment: Five Panzers and one truck. Three **Pz.38(t)** had to be blown up because they broke down, one **Pz.38(t)** fell into a tank trap and couldn't be recovered in spite of much effort, and one **Pz.38(t)** ran onto a mine.*

Combat report of a **Panzer-Nachspitze** (rear guard) from 15 to 24 January 1942

*15Jan42 - An attack group led by Hptm. Kahl consisting of five **Pz.Kpfw.38(t)**, one **Pz.Kpfw.IV**, one **Panzerjaeger-Zug** of two **5 cm** and one **3.7 cm Pak** was sent to cooperate with **Infanterie-Regiment** 215 to guard the south flank of IX Armee Korps.*

*16Jan42 - **Panzer-Nachspitzen** (rear guards) were created and assigned to the individual divisions to cooperate with the division's rear guards in covering the retirement of the IX Korps to a position east of Gschatsk.*

*Strength of the **Panzer-Nachspitzen** was:*
Two **Pz.IV**, three **Pz.38(t)**, two **5 cm Pak** led by Oblt. Kastorff with the **87.Inf.Div.**
Four **Pz.38(t)**, one **5 cm Pak**, two **3.7 cm Pak** led by Lt. Schlueter with the **78.Inf.Div.**
Five **Pz.38(t)** and two **5 cm Pak** led by Oblt. von Harnack with the **252.Inf.Div.**

*Losses were four **Pz.38(t)** due to engine failure which because of time constraints couldn't be recovered and were blown up.*

20.Panzer-Division Pz.Kpfw.38(t) Operational Status Reports				
Date	Total Available	Operational	In Repair	Total Loss
1Mar42	20	12	8	
30Apr42	26			
20Jun42	37	30	7	
31Jul42	38	32	6	
26Aug42	28	15	13	14
30Sep42	34	30	4	4
1Nov42	35	25	6	
28Nov42	34	31	3	
31Dec42	27	15	12	11
31Mar43	13	11	2	
5Jul43	9	8	1	

On 1 March 1942, the **20.Pz.Div.** reported that they still had 20 **Pz.Kpfw.38(t)**, of which 12 were operational. They had a total inventory of 26 **Pz.Kpfw.38(t)** on 30 April 1942, having received 50 replacements since 22 June 1941 and total losses of 142. On 13 June 1942, **Panzer-Regiment 21** reported that another 15 **Pz.Kpfw.38(t)** replacements had been promised and were on the way. On 31 July, the **III.Abt./Pz.Rgt.21** was active with the **1.**, **9.**, and **10.Pz.Kp**. The last 10 **Pz.Kpfw.38(t)** sent as replacements were received by 16 September 1942. On 1 November 1942, there were 6 **Pz.Kpfw.38(t)** with the **III.Abt. Stab**, 10 with the **9.Pz.Kp.**, 9 with the **10.Pz.Kp.**, and 4 with the **2.Pz.Jg.Kp.** of the **Pz.Jg.Abt.** There were still 6 **Pz.Kpfw.38(t)** with **Pz.Rgt.21 Stab**, 6 with **III.Abt.Stab**, 11 with the **10.Pz.Kp.** and 8 with the **2.Pz.Jg.Kp.** on 1 February 1943. But in March and April there were only 7 **Pz.Kpfw.38(t)** reported to be in use with the **III.Abt.Stab**. There were still 9 **Pz.Kpfw.38(t),** of which 8 were operational directly before the start of Operation Zitadella on 5 July 1943.

22.Panzer-Division with Panzer-Regiment 204

Panzer-Regiment 204 (created with the **I.** and **II.Abt.** and **1.** to **6.Pz.Kp.** on 7 July 1941) was assigned to the **22.Panzer-Division** by orders dated 25 September 1941. As ordered on 7 February 1942, the **22.Panzer-Division** was to be transferred to Heeresgruppe Sued starting on 17 February. On 3 March 1942, the **III.Abt./Pz.Rgt.204** was ordered to be immediately formed with the **7.** and **8.le.Kp.** organized in accordance with **K.St.N.1171** dated 1 November 1941 and outfitted with **Pz.Kpfw.38(t)** to be combat ready by 14 April 1942.

Panzer-Regiment 204 was sent in with the **22.Panzer-Division** in the last major attack of a Panzer-Regiment fully outfitted with 79 **Pz.Kpfw.38(t)** and **Pz.Bef.Wg.38(t),** as related in the following report written by the commander of **Panzer-Regiment 204**, Oberst Koppenburg, about the attack on Korpetsch on 20 March 1942:

The regiment's advance occurred without incident from the area of Stary Krim to the rest area in a fruit plantation south of Bairatsch.

The further advance from there to the staging area north of Wladislawowka was started at around 2130 hours during a very dark night. The I.Abteilung arrived at the staging area about 0200 hours. The Regiment Stab and II.Abteilung encountered major difficulties during the advance caused by motorized columns and horse-drawn vehicles in both directions. They especially delayed a smooth passage at bridges. The number of men posted for directional control on the previous comparatively bright nights were insufficient on this night, so that the Regiment Stab and II.Abteilung arrived at the staging area between 0415 and 0500 hours. At 0445, the SPW-Kompanie arrived. Soon thereafter the Fla-Kompanie also reported in. Directly before leaving the staging area, the 7.Kompanie/Schuetzen-Regiment 140 reported in to be transported on the Panzers of the II.Abteilung. The formation of the reinforced Panzer-Regiment in the staging area is shown on the attached map. Contact of I.Abteilung with the SPW-Kompanie was ensured by a meeting between both commanders. The SPW-Kompanie did not have any radio sets that were of a frequency that could be used to communicate with the I.Abteilung or the Regiment. Also, there was no contact with the Fla-Kompanie, so that long-term combat orders had to be given to them. Nothing could be seen of the Schuetzen-Regiment located on the right flank. At 0455 hours, dawn broke. The light barely penetrated the thick fog. At this time visibility was limited to 50 to 100 meters.

The order to advance was given by me at 0500 hours. The I.Abteilung started; the SPW-Kompanie remained stationary and thereby delayed the start of the II.Abteilung so that too large a gap occurred for the Abteilung to maintain sight contact in the limited visibility. The first 1000 to 1500 meters advance toward the northeast crossed an old circular defense position, that on my orders was crossed very slowly because of the limited visibility. I then ordered a halt from 0515 to 0530 hours, but the fog didn't clear. Elements of the Schuetzen-Brigade arrived from the south, and a short meeting was held between both regimental commanders and the commander of the I.Abteilung.

At 0530 hours, the advance restarted with the motorized infantry. The SPW-Kompanie hadn't made contact with the I.Abteilung, but found itself near the II.Abteilung. From here on, no sight contact (only radio contact) occurred between me and the I.Abteilung and between the I. and II.Abteilung. Therefore the further development of the battle had to be handled as two separate attacks.

Abteilung Collin *apparently turned too far to the north. From later events it was learned that they had crossed our forward lines at and east of Point 26.7. I traveled at the start with the II.Abteilung. That the I.Abteilung drifted to the left was unknown to me.*

*About 0600 hours, the **I.Abteilung** had arrived at a position on the track from Korpetsch to Point 26.7 when 25 to 35 enemy light tanks attacked from the front through the fog. The **I.Abteilung** took up the firefight and most of the enemy tanks turned away. The **Abteilung** was formed on a wide front and wanted to start the chase when from almost the same direction about 10 KW and T 34 tanks advanced, escorted by light tanks. Strong artillery fire hit the position of the **I.Abteilung** at the same time as the first hits from Russian 7.62 cm tank guns. Then the **2.Kompanie** and the left wing of the **3.Kompanie** ran into an anti-tank minefield that hadn't been cleared. The individual Panzers could have been warned in time by our infantry occupying this position.*

*The **I.Abteilung** lay under strong enemy fire with visibility hindered by the smoke and fog. They received many hits from the 7.62 cm tank guns fired at long range. The **1.** and **2.leichten Kompanien** on the wings were pulled back somewhat. The **3.mittlere Kompanie** continued the firefight. Then the commander pulled his **Abteilung** back to in and behind our forward defense lines under the protection of anti-tank guns located there. The anti-tank guns falsely interpreted the 200 to 300 meter rearward movement of the Panzers, hooked up, and drove away to the rear.*

*Toward 0630 hours, new enemy tanks appeared on the right flank, causing the right wing of the **I.Abteilung** to bend back. Under the still heavy effective enemy artillery fire, mainly from the north and apparently also from the northeast, the **I.Abteilung** was forced to decide to pull back farther to obtain visibility by getting out of the thick smoke and wall of fog.*

*I was continuously notified by radio of the fighting of the **I.Abteilung**. However, I thought that, as planned, the **I.Abteilung** was located in front of me in the area west of Korpetsch. In response to the urgent calls for help from the **I.Abteilung**, I ordered the **6.mittlere Kompanie** to support the **I.Abteilung**, which naturally couldn't find them. At the same time, I requested fire support to fall on the area northwest of Korpetsch from the artillery commander with whom I was in radio contact. I also requested that a battery be pulled forward to directly fire at the KW tanks. At this time I was located in the area west of Korpetsch.*

*On the report from the **I.Abteilung** commander that he couldn't hold any longer against the superior forces and had taken 30 to 50 percent losses, I concurred with his proposal, while fighting to slowly pull back toward the **II.Abteilung** with the understanding that I hoped thereby to personally regain contact with the **I.Abteilung**. The **I.Abteilung** then slowly fell back, still fighting with the enemy tanks. My radio contact with the **I.Abteilung** was sometimes interrupted. About 1000*

Above: A Pz.Kpfw.38(t) and Pz.Kpfw.II in the 7.Kp./Pz.Rgt.204 with the 22.Pz.Div. in 1942. (KHM)

hours, the **I.Abteilung** arrived in the area northeast of Wladislawowka and gathered there. I learned of this position of the **I.Abteilung** by a roundabout way through the division radio report at about 0930 to 0945 hours. At the same time, the strength of the **I.Abteilung** was reported as 41 Panzers. I couldn't explain this contradiction with the earlier report.

At about 0630 hours, the forward elements of the **II.Abteilung** arrived in the area west of Korpetsch. They advanced in the general direction of Point 28.2, apparently passed to the left of a Russian minefield, and reached the stream bed from Korpetsch to Tulumtschak. Part of the **SPW-Kompanie** wound up in the minefield and part went forward with the front half of the **II.Abteilung**. The **7.Kompanie/Schuetzen-Regiment 140** dismounted from the Panzers and took up positions in the old German or Russian field works west of Korpetsch. Russian infantry retired to the north and northeast in front of the **II.Abteilung**. The enemy infantry, taking heavy losses from tank machine gun fire, were reduced to a small element that remained sitting in the foxholes, fought concealed, and raised their hands high and then sometimes shot from behind at the Panzers and the commanders who from time to time looked out of the open hatches.

When the area west of Korpetsch was reached, strong enemy artillery barrages struck the area. This was fired mainly by the batteries at Dshantora but a few also from the northeast and east. The Russian artillery fire followed all movements of the Panzers with outstanding flexibility, so that the Panzers were forced to continuously change their positions in all directions in a depth of about 1.5 to 2 kilometers on both sides of the track from Korpetsch to Point 26.7.

Toward 0700 hours, the commander of the **II.Abteilung** arrived at the stream northwest of Korpetsch. Reconnaissance along the stream by the light platoons revealed that this was completely dug out as an anti-tank ditch. The steeply protruding walls on the southwest bank created an absolute barrier to Panzers. The information received on the day before the attack that it was possible to cross with a **Kuebelwagen** is unexplainable. Anyway, the **II.Abteilung** attempt to cross the anti-tank ditch resulted in a considerable number of Panzers remaining stuck.

During this entire time, with a few pauses, the enemy artillery fire held up with undiminished ferocity. Considerable damage to the Panzers was caused by direct hits and fragments that partially immobilized, partially jammed the turrets, and damaged the weapons.

Because of the strong artillery fire, the **Regiment Stab** and **II.Abteilung** spread out both in width and depth to weaken the fire concentration as much as possible. About 0730 to 0800 hours, enemy guns and advancing anti-tank guns were identified on the left flank on the path from Korpetsch to Tulumtschak and engaged.

At the same time, from the area northwest of Point 28.2, a strong, very wide enemy infantry attack approached toward Korpetsch and to the west. The attack broke down under the fire of the Panzers located by the anti-tank ditch.

At the same time, the battle proceeded against the stationary enemy infantry and field works west of Korpetsch and northeast of the stream. In spite of the thinning fog, visibility was strongly hindered by the continuous artillery bombardment.

Upon the order for the **6.Kompanie** on the right wing to go support the **I.Abteilung**, the commander of the **6.Kompanie** pulled out to the right, managing to cross the anti-tank ditch at a location northeast of Korpetsch. Three **Pz.Kpfw.IV** managed to cross this barrier, while the fourth and a **Pz.Kpfw.II** remained stuck at this location and blocked the crossing point. Oberleutnant Luckhardt, the commander of the **6.Kompanie**, went forward to Point 28.2 and reached this height at about 0730 hours. A report about this did not reach either the **Abteilung** commander or me because the antenna on the commander's Panzer was shot off. The Panzers at Point 28.2 were identified on the height, partially not as ours but as the enemy, while Major Urban himself held that these were Panzers from the **I.Abteilung**. Regrettably, they were shot at by Panzers on this side of the stream; however, the fire was halted by Major Urban. Along the way Oberleutnant Luckhardt destroyed elements of the Russian occupants on the north edge of Korpetsch, many guns and anti-tank guns, and successfully engaged observation posts at Point 28.2 and enemy infantry that were pulling back.

At approximately the same time that Oberleutnant Luckhardt reached Point 28.2, about 0830 hours, an enemy infantry attack from Tulumtschak in the left flank was recognized and engaged by the rear half of the **II.Abteilung** (where I was located at this time), brought to a halt, and then forced to retreat.

Since there was no chance of crossing the anti-tank ditch under the well-placed enemy artillery fire with our own equipment without support from infantry and pioneers, with my concurrence the **II.Abteilung** was pulled back slowly, taking along the wounded and crews that had abandoned their Panzers. Slowly retiring to the south, the **II.Abteilung** arrived at the railway embankment directly west of Korpetsch. I saw nothing of the retirement of this **Abteilung** through the prevailing smoke and because of my observation to the north and northwest in the direction of Tulumtschak. From about 0915 hours on, there was no longer any radio contact with the commander of the **II.Abteilung**, who because of the loss of his Panzer's radio, wanted to change to another.

Because I had to calculate from the orders given to the **I.** and **II.Abteilung** that as a result of the ordered rearward movement, these must run into me and the Panzers around me in the area west and northwest of Korpetsch, I waited there with about 15 Panzers from the **II.Abteilung** and the **Regiment Stab** until 1010 hours.

In the meantime, the report of the new location of the **I.Abteilung** came in through the division. The report on the retirement of the forward elements of the **II.Abteilung** was brought over by the ordnance officer driving in a Panzer from the **II.Abteilung**.

In addition, at about 1000 hours, Oberleutnant Luckhardt appeared in the Panzer that had been used by the doctor in the **II.Abteilung**, and reported to me that, being shot up, his three *Pz.Kpfw.IV* had to be abandoned at Point 28.2, that he himself came back on foot, and that he didn't know anything specific about the fate of both other crews.

At this time, various reports came in at the same time from several Panzers that their fuel and ammunition were almost exhausted. Because contact had still not been made with the infantry, after reporting to the division I decided to pull back from here with the remaining elements of the Regiment to regain control of the Regiment at the assembly area of the **I.Abteilung**.

I myself could only observe rearward movements on this part of the front. I believe that they went from the rear to the front. During the retirement of the Panzers from the artillery fire that lasted two to three hours, apparently the impression was awakened in several rearward elements that the Panzers wanted to pull back. I had the impression that the wave of panic passed to the elements farthest back in the area of the railway embankment, infantry as well as motorized elements (**SPW**, Panzer, self-propelled Flak). When I recognized this at about 0915 hours, I sent the adjutant of the Regiment back in a Panzer to stop the retreat and turn them back to advance in the direction of the attack. This was done with some success.

Of the 33 Panzers that were left lying in front of our lines:
o Two *Pz.Kpfw.II*, six *Pz.Kpfw.38(t)*, one *Pz.Kpfw.IV* were burned out or so damaged by hits that their repair doesn't appear possible.
o Five *Pz.Kpfw.II*, ten *Pz.Kpfw.38(t)*, and three *Pz.Kpfw.IV* were so heavily damaged that by cannibalization of parts, other Panzers can be repaired and made operational.
o Three *Pz.Kpfw.II*, one *Pz.Kpfw.38(t)*, and two *Pz.Kpfw.IV* have little damage or are stuck and potentially can be made operational.

Because we already managed to tow away many Panzers during the day of the attack, and from the situation it was calculated that the main part of the Panzers lying between and behind the lines could still be recovered by a renewed attack for which the Regiment assembled, an order to destroy the Panzers was not given.

Attempts during the following nights to recover the Panzers failed because of the enemy defense.

The attack failed because of inadequate preparations and coordination. **Panzer-Regiment 204** had started with 45 **Pz.Kpfw.II**, 77 **Pz.Kpfw.38(t)** (including **Pz.Bef.Wg.**), and 20 **Pz.Kpfw.IV**. The final tally of total writeoffs from the action on 20 March was 9 **Pz.Kpfw.II**, 17 **Pz.Kpfw.38(t)**, and 6 **Pz.Kpfw.IV**.

Panzer-Regiment 204 was sent 30 **Pz.Kpfw.38(t)** replacements that arrived by 30 April 1942. On 1 May 1942, **Panzer-Regiment 204** reported that there were 5 **Pz.Kpfw.38(t)** in the **Regt.Stab**, 39 with the **I.Abt.**, 39 with the **II.Abt.**, and 37 with the **III.Abt.**

The **III.Abt./Pz.Rgt.204**, sent into action with **Panzer-Brigade 22** with A.O.K.11 in the Crimea, reported having 21 **Pz.Kpfw.38(t)** operational on 7 June, 17 on 23 June, 23 on 25 July, and 10 on 18 August.

The rest of **Panzer-Regiment 204** had been issued 20 **Pz.Kpfw.III** with **5 cm Kw.K. L/60** and therefore took only 50 **Pz.Kpfw.38(t)** into action with them at the start of the Summer offensive on 7 July 1942 and were down to 28 operational on 28 July, 18 on 25 July, 18 on 2 August, 34 on 11 August, and 10 on 24 August. There were 36 operational **Pz.Kpfw.38(t)** with **Panzer-Regiment 204** on 20 September, with another 79 **Pz.Kpfw.38(t)** that hadn't been manned with crews. From a high of 79 **Pz.Kpfw.38(t)** reported operational on 10 October, after giving away 51 **Pz.Kpfw.38(t)** to **Panzer-Verband 700**, **Panzer-Regiment 204** was down to 25 operational on 31 October, 26 on 10 November, but only 5 on 18 November when the Russians attacked to surround Stalingrad. Only 1 **Pz.Kpfw.38(t)** was reported to be operational with **Panzer-Regiment 204** from 20 November to 1 December 1942. The **1.**, **3.**, **4.**, and **6.Pz.Kp./Pz.Rgt.204** with the **22.Panzer-Division** were written off on 15 to 20 March 1943.

27.Panzer-Division with Panzer-Abteilung 127

On orders from Hitler, part of the **22.Panzer-Division** detached as **Gruppe Michalik** was renamed the **27.Panzer-Division** on 31 August 1942. As ordered on 20 September, the **III.Abt./Panzer-Regiment 204** was renamed **Panzer-Abteilung 127** (effective 1 October 1942), and the **7.** and **8.le.Pz.Kp.** and **9.m.Pz.Kp./Pz.Rgt.204** were renamed **1.**, **2.**, and **4.Kp./Pz.Abt.127**. Out of a total authorized strength of 35 **Pz.Kpfw.38(t)** they possessed a total of 24 **Pz.Kpfw.38(t)** and 2 **Pz.Bef.Wg.38(t)** of which 16 were operational on 20 August 1942. Reporting one total loss on 29 September and three (**Fgst.Nr.1279, 1333, 1344**) returned to the Heereszeugamt on 10 September, **Pz.Abt.127** had 22 **Pz.Kpfw.38(t)**, with **Fgst.Nr.795, 798, 821, 823, 825, 835, 858, 917, 1050, 1140, 1252, 1277, 1280, 1286, 1320, 1321, 1327, 1330, 1331, 1332, 1335,** and **1336,** of which 22 were operational on 30 September, but only 4 were operational on 1 October and 6 on 4 October, 5 on 11 October, 5 on 21 October, 2 on 1 November, 5 on 20 November, 4 on 11 December, 0 on 20 December, 2 on 1 January, 2 on 11 January, 1 on 21 January 1943, and 1 on 10 February 1943. On orders dated 15 February 1943, the **27.Panzer-Division** was disbanded and the remnants incorporated into the **7.Panzer-Division**.

Panzer-Verband 700

On 16 October 1942, OKH Org.Abt. ordered Heeresgruppe B to create **Panzer-Verband 700** consisting of an **Abt.Stab u. Stabs.Kp.**, 1 **Pz.Sp.Kp.**, 3 **le.Pz.Kp.(t)**, 1 **m.Pz.Kp.(deutsch)** that was to be incorporated into the **16.Inf.Div.(mot)** as soon as possible. **Pz.Rgt.27** in the **19.Pz.Div.** was to give up one **le.Pz.Kp.(t)** without Panzers, and **Pz.Rgt.204** in the **22.Pz.Div.** was to provide 2 **le.Pz.Kp.(t)** with Panzers and a **m.Pz.Kp.** without Panzers. The **Stab u.StabsKp.** was to come from the **III./Pz.Rgt.36** in the **14.Pz.Div.** The unit was to be outfitted with excess **Pz.Kpfw.38(t)** from the **22.Pz.Div.** and newly issued **Pz.Kpfw.IV** for the **4.m.Pz.Kp.**

A status report on **Panzer-Verband 700** dated 5 November 1942 revealed that it had 0 **Pz.Kpfw.38(t)** in the **1.Kp.**, 26 in the **2.Kp.**, and 25 in the **3.Kp.**, of which 45 were operational. As revealed by their **Fgst. Nr.** (499, 581, 608, 654, 655, 726, 728, 777, 778, 788, 791, 802, 813, 824, 859, 923, 925, 932, 937, 939, 941, 943, 947, 1044, 1183, 1267, 1271, 1284, 1287, 1290, 1323, 1341, 1342, 1351, 1355, 1356, 1488, 1489, 1512, 1513, 1515, 1516, 1517, 1519, 1521, 1523, 1524, 1525, 1526) these were a mixture of **Pz.Kpfw.38(t) Ausf. E, F, S,** and **G,** including the last **Ausf.G** completed in June 1942.

In a status report dated 23 November 1942, **Panzer-Verband 700** revealed:

*Two **leichte Panzer-Kompanien** and a **Werkstattzug** (maintenance platoon) from the **22.Panzer-Division** and one **leichte Panzer-Kompanie** from the **19.Panzer-Division** have already arrived. The Panzer for these three **leichte Kompanien** are in very poor condition. It must be reckoned that 1/3rd will continuously be in repair because of mechanical problems.*

*Because of their armor and armament, these Panzer can't be used to combat Russian tanks. The **Typ 38(t)** can only be referred to as totally outdated. It is also mechanically unsuitable for the planned utilization in desert areas. At external temperatures over 30 degrees C, the engine overheats and stops. To fulfill the assigned objective, the unit must be outfitted with at least **Pz.Kpfw.III (lang)**.*

On 7 February 1943, the remaining elements of **Panzer-Verband 700** in the A.O.K.2 sector were ordered to be incorporated in the **4.Pz.Div.** under A.O.K.4.

Panzer-Verband 700 Pz.Kpfw.38(t) Operational Status Reports				
Date	Total Available	Operational	In Repair	Total Loss
5Nov42	51	45	6	
27Nov42	50	17	33	1
6Dec42	50	26	24	
16Dec42	50	25	25	
6Jan43	45	19	26	5
13Jan43	27	4	13	18

Directly before the start of Operation Zitadelle on 30 June 1943, there were only 13 **Pz.Kpfw.38(t)** left with German units on the Eastern Front, of which 3 were with the **8.Pz.Div.** and 9 with the **20.Pz.Div.**